The Power of Writing Well

Why and How to Craft Business and Technical Communications for:

WEALTH • HAPPINESS • LEADERSHIP

by

PETE GEISSLER

Printed in the United States of America

First Edition, 2013

ISBN 1-933704-06-3

The Expressive Press
www.TheExpressivePress.com

CONTENTS

PART I

PART II & WORKBOOK

PART I

FOREWORD, CAVEATS, AND TIPS FOR CREATING YOUR COMPETITIVE ADVANTAGE

I wrote *THE POWER OF WRITNG WELL* to address everything managers, engineers, and scientists need to be better senders and receivers, not to cover everything they need to know about the language or to be the perfect sender or receiver; nobody is.

Pete 6/13/13 1:38 PM
Comment: *Sender* is my catchall word for writer and speaker, and *receiver* for reader and listener. Both words are used as singular or plural nouns.

The many books on writing and communicating that claim to be everything to everybody fail simply because they are overwhelmingly complex, full of jargon and useless labels and distinctions such as participial phrase as opposed to gerund phrase, or transitive verb versus intransitive. Most of us outside the Ivory Towers of academe don't care, and we shouldn't since they are not relevant to our needs.

This short book is a synopsis of habits and techniques—your tools-- that work most of the time for most of the people who write at work and want to be happier in all parts of their lives: nothing more, nothing less. It is also a true and accurate reflection of my forty years of writing for business and of teaching writing at two prestigious universities and many professional societies and companies. You can trust that what I'm telling you will improve your abilities to communicate and think, and make you more productive, promotable, and happy. It will also make your organization more efficient and profitable.

Pete 6/13/13 1:38 PM
Comment: My definition of *concise*, as you'll learn later.

I guarantee it, and my students attest to it.

Pete 6/13/13 1:38 PM
Comment: Your first exposure to active voice, an important tool. I'm the actor in the first phrase, my students in the second.

The book is also a reflection of *how* I've taught and still teach, which I think is unusually effective. Basically, I'll introduce a useful tool in abstract terms, and then illustrate it with examples, often with before-and-after examples to demonstrate the transformation from bad to good. In that way, I'll put the tool into the real world: yours, where it belongs. Then I'll comment as you read, and then summarize with more before-and-after examples, some of which I hope you'll resonate with—Aha; I've read or written or heard that!-- and be amused by.

Pete 6/13/13 1:38 PM
Comment: I show how to write better more than tell.

Pete 6/13/13 1:38 PM
Comment: I'll comment within the text and in balloons to illustrate both methods.

The before examples—correctable or improvable— are always italicized; the after—corrected or improved—in regular type.

Pete 6/13/13 1:38 PM
Comment: A form of parallelism, as you'll see later.

YOU CAN CREATE YOUR PERSONAL COMPETITIVE ADVANTAGE by understanding your incentives and adhering to all the tools, connecting them in your mind to create a cohesive whole, applying them in the more complex before-and-after examples in Chapter VII (and, if you're so inclined, in APPENDIX VII) and applying them to your everyday writing immediately.

The sequence from tools to simple examples to more complex examples will, I guarantee you, unleash the power in your words.

Enjoy, learn, profit.

Pete Geissler

CHAPTER I: UNDERSTAND YOUR INCENTIVES. WHY WRITE? WHY WRITE BETTER? THE ANSWERS WILL SURPRISE YOU

Writing is the basis for speaking and thinking; therefore, it is the basis for wealth, happiness, and leadership.

Writing is far more than words on paper or in your computer; it is a permanent reflection of your thinking, your intelligence, your logic, and your abilities to explain complex ideas to yourself and to others. As such, it determines your future in commerce probably more than any other skill. Think of your writing this way: muddy thinking yields muddy writing, and vice versa: Muddy writing displays muddy thinking. You don't need that in this age of increasingly complex information.

> Pete 6/13/13 1:38 PM
> **Comment:** Yes, writing is probably the best way to make sense of your world, as well as the worlds of others.

Writing better than your peers puts you in the lead for promotion or for success as an entrepreneur, as explained below. It also makes you a better speaker, your more public display of your intelligence.
The rewards far outstrip the efforts, and they permeate every part of your life. If you aren't convinced of that, please read my book, *The Power of Being Articulate.*

> Pete 6/13/13 1:38 PM
> **Comment:** It also makes you happier in your social life via clear thinking that leads to better, more constructive decisions. See FURTHER READING at the end of this book.

To be more specific, good writers are:

1. **More productive** simply because they know and can expeditiously apply the techniques of good sending and receiving, and, therefore, more quickly become practitioners. Productivity continues to rise for these folks in a never-ending upward spiral as they realize that their sending and receiving are improving. Receivers realize it too and compliment the senders, and senders become more confident of their abilities to express their thoughts and *want to send more.* One inevitable result: practitioners become more valuable contributors to their own wealth and happiness and to the success of any organization, and are ...

> Pete 6/13/13 1:38 PM
> **Comment:** Notice the use of heads –bold face type—that run into the normal text and add greatly to concision; we'll delve later into heads of various types. BTW, some writers eschew heads for 'headings' or 'headlines'. Your choice.

2. **More promotable** since they display their intelligence via their words. Most of us automatically label folks who mangle the language as "not too bright", and label those who express themselves clearly and concisely as "impressive" or "brilliant".
   NOTE: The Wall Street Journal, in a special issue of September 9, 2002, reported that "...employees may get by on their technical and quantitative skills for the first few years out of school, but soon leadership and communications skills come to the fore in distinguishing managers whose careers really take off."
   Good practitioners also tend to be ...

> Pete 7/23/13 5:23 AM
> **Comment:** The marketing manager for a large manufacturer responded to a letter asking for more information concerning a proposal in a parallel, cohesive format (see below) and so clearly and concisely that his firm was awarded the contract at a premium price. He was promoted to Division Manager, where he immediately insisted that all project, sales, and marketing employees attend courses in writing.

> Pete 6/13/13 1:38 PM
> **Comment:** Think of George W. Bush and the resumes that I get from folks who want to be professional writers; If they are well-written, I pass them on to colleagues who are hiring; if not, I pass them on to the dustbin.

3. **More creative** since most of us think more in words and less in numbers or pictures. We connect known concepts in words to create new concepts—the essence of creativity—and the more

words we know the meanings of and can use the more creative we can be. Practitioners also are ...

Pete 6/14/13 5:43 AM
Comment: E. M. Forster said it best: 'How do I know what I think until I see what I write.'

4. **More respected** since they exploit the techniques of writing to cleanse their thoughts of irrelevancies; they are insightful and it shows in their words.

The personal benefits of good writing trickle up, down, and sideways in every organization from families to businesses to governments to create tangible benefits that accrue to everyone involved. They are:

1. **More profitable and efficient** via higher productivity of employees/members.
2. **More competitive and sustainable via** more creative responses to requests from clients, members, and associates.
3. **More directed via clear leadership,** as I point out later.

The bottom line is perhaps expressed most succinctly by John Yasinsky, former CEO of GenCorp.: " I am totally convinced that articulation, expressed most often in good writing, is the tiebreaker in everyone's professional, financial, and personal lives." [ii] In other words, they are wealthier, happier, and better leaders; their organizations are more productive, profitable, and their brands more identifiable.

Pete 6/13/13 1:38 PM
Comment: John's thoughts on the importance of being articulate are expressed more fully in my book, *The Power of Being Articulate*.

[i] *The Power of Being Articulate,* by Pete Geissler, Kallisti Press, Wilkes Barre, PA, 2013

You, the Sender/Receiver

Better writer/listener

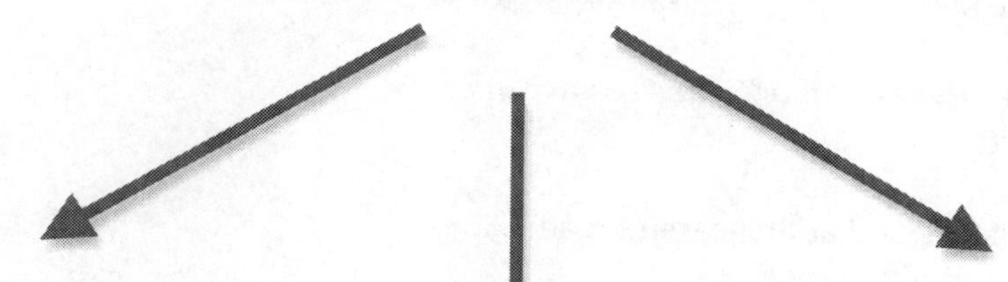

More promotable

--writing reflects your intelligence, so …

--know your subject and what you want to say about it

--say it clearly, concisely, logically, on-point

More productive

--you

--your receivers

--your firm, organization, family

More profitable/efficient

--you

--your firm, organization, family

CHAPTER II: LIVE THE EIGHT EVERYDAY HABITS OF SUCCESSFUL SENDERS

Pete 6/13/13 1:38 PM
Comment: *It bears repeating: Senders* is my catchall for writers and speakers, and *sendings* for documents and spoken words.

1. ***Become an analytical receiver; when you do, you will be a more analytical sender***

No less a writer than Ernest Hemingway said that the first prerequisite to being a writer is to develop a built-in, infallible, s**t detector—which literally demands that you be a careful, attentive, empathic listener. I start you on that road with my before-and-after examples, and then you must carry the ball. Edit everything that you hear and read, no matter how trivial, to develop the habit. Then edit everything that you write and say, and watch your sendings of all types improve. It's fun and profitable; trust me. You can start with this book; surely you can find places to improve it and you'll let me know.

Pete 6/13/13 1:38 PM
Comment: You wouldn't want me to sanitize his words, would you? Wouldn't that blunt his sword by being out of character?

2. ***Always meet the two definitions of 'good'; they're your goals***

Good writing is defined by its characteristics. It is:

13. **Clea*r:*** your words are understood at first careful reading or hearing; they cannot be misunderstood.
14. **Concise**: you've used only the words and thoughts needed to meet your purpose(s), and those of your receivers. Not too brief, not too wordy; either causes the tag games that erode productivity.
15. **Purposeful**, aka on-point or focused: Every sending has a job to do, and it is always to influence receivers to act or behave in new ways, usually in ways that benefit senders and receivers. In non-fiction, it's always to buy your gizmo or service; in fiction, it's to buy your book or story to entertain; in politics (a form of fiction) it's always to be elected. I guarantee that knowing your purpose will create a sending that gallops to its point most quickly and dramatically (your 'hook') and will keep you on-point; you will be most clear on what is important and what isn't. See Appendices I and II for more on this imperative.

Pete 6/13/13 1:38 PM
Comment: I'm referring here to sendings in business, one type of non-fiction.

Good writing is defined by its results. It:

1. **Raises profitability** by one to ten percent of sales, based on my informal survey of some twenty marketing and top managers. One said that if he could magically convert all the bad writing his employees create his profits would double; another said that he fears that bad writing will lose so many customers that his firm will go out of business. The following four bullets support their statements; good writing:

- **Raises the productivity of individuals and firms.** Receivers—whether customers or employees-- understand the message at first reading; they do not need to ask senders for clarification, avoiding the irritating and time-consuming tag games that erode productivity.
- **Increases management efficiency**. Executives trust that the writing of others is clear, concise, on-point, and sensitive to the needs and wants of receivers . As a result, they are freed for other tasks for which they are more suited. Costs plummet and schedules accelerate.
- **Retains and creates customers**. Customers are impressed with senders' intelligence, raising hit rates and lowering sales costs. Well-written proposals are easier for customers to evaluate and will get the initial order, and well-written reports will get repeat business--assuming other evaluation factors are similar. See the anecdote about the marketing manager above.
- **Creates competitive advantage.** Higher productivity, lower costs, clearer expressions of your intelligence—all add up to serious and sustainable competitive advantage for individuals and firms. Please see Appendix VII for details.

Pete 6/13/13 1:38 PM
Comment: I learned a valuable lesson about the sensitivities of receivers when I was in college .I wrote a letter to a close friend in a very formal tone that I was practicing. She was offended to the point where she cut off all communications with me for years. I'm not sure she ever forgave me; I know that I never forgave myself.

Pete 7/23/13 5:31 AM
Comment: One of my students told me that she was awarded scholarships totaling in six figures largely because her proposals were written so well. She continues: "I could not have attended graduate school without that money."

3. ***Fantasize to empathize****; only then can you know your receivers*

Pete 6/13/13 1:38 PM
Comment: It deserves repeating: : *Receivers* is my catchall for readers and listeners.

You can't know how receivers will react to your words until you tap into your distinctly human talent to infer the mental states and reactions of others by leaning back and fantasizing. For help, complete the Receivers' Profile in Appendix I, then contemplate how your answers will shape your sending's content (what you say), structure (in what order you say it), and tone (your choice of words).

Pete 6/13/13 1:38 PM
Comment: Call it focused daydreaming if you want. Regardless, it's a route to on-point sendings and creativity.

Tone can be described by comparing your careful choice of words when writing or speaking to a person of authority such as your boss or a judge (creating a formal tone) with your more casual choices when writing and speaking to your closest friend (creating a conversational tone).

Pete 1/8/14 3:26 PM
Comment: I can tell many stories of how not knowing receivers has cost big bucks. Here's one: I was sent to a conference on acid rain and to write articles on what I heard. Two reps of a pump manufacturer talked at length about the wonders of their pump's design and metallurgy; receivers, all plant engineers, wanted to hear about maintenance, capacities, and efficiencies, and were fidgeting only a few minutes into the presentation. The manufacturer wasted thousands of dollars.

When you profile your receivers, always include lawyers and judges. They are rarely if ever your primary or secondary receivers, but they can react to your sendings in profound and ugly ways should your sendings ever cause a lawsuit that is likely based on ambiguity in a proposal or report.

4. ***Start early; your subconscious smarts will kick in and do the hard work***

Your subconscious mind is the root of the great ideas that help to separate good from bad sendings, but only if you give it time to do so and you exploit your recorder to its fullest.

You need time to "cook" your ideas, i.e. to fully develop them by answering *why* your ideas are good or bad, and *how* you know. Ask yourself why and how seven times before you are satisfied that your ideas are fully developed.

Your recorder—there's one on your smart phone-- is your most valuable bed and bath partner, so put one within easy reach and spit out those great ideas that come to you in the middle of the night or while taking a shower. Carry your recorder when walking the dog, driving—doing anything during which your mind isn't fully occupied. Use it to record thoughts after a meeting, an idea from a conversation, TV show. Whatever.

Pete 6/13/13 1:38 PM
Comment: Many writers keep a log/notebook in which to jot down ideas as they emerge from their minds. My log is my recorder.

Stuff your mind with your writing—or any creative-- dilemma before going to bed and watch the dilemma go away.

Pete 7/23/13 5:35 AM
Comment: When I needed a great theme for a speech I was writing for an executive, I stuffed my mind with everything I knew about the subject and receivers and went to bed. I awoke a few hours later with the germ of a theme that I immediately recorded, then fleshed out the idea first thing in the morning.

I absolutely guarantee that this one habit will raise your productivity and the quality of your sendings.

Pete 6/13/13 1:38 PM
Comment: Note the active voice that I discuss in Chapter IV. I could have written: Your productivity and quality of your sendings will improve if you start early. Who says so?

5. ***Don't be smitten with what you've written; if you are, you can't edit to improve***

Unless you're a genius, your first draft(s) won't be "good" as I've defined it. Even Shakespeare needed to rewrite often to get it right, and Stephen King cuts and revises his words brutally. I may call the draft I submit to my clients/editors my first so that they are more comfortable with their critique—they know I expect it-- but they are really my fifth or tenth. They don't need to know that.

Pete 6/13/13 1:38 PM
Comment: King's book *On Writing* is marvelous; I suggest that you read it.

Pete 6/13/13 1:38 PM
Comment: You are reading my fifth draft at least; I lost count long ago, and I am a pro writer wallowing in a topic that I know well and is dear to me.

Editing yourself is your toughest job made easier if you follow one rule: delete everything that does not contribute to the story line, everything that does not further your or your receivers' purpose(s).

6. ***Be a slave to the only writing process that works; you'll be amazed at how it helps***

- **Think before you write**

Pete 6/13/13 1:38 PM
Comment: Remember, your writing reflects your thinking, aka your intelligence. If you agree, it makes sense to think carefully about your writing before you start, during the actual crafting of your words, and after receivers react.

❖ Complete the Receivers' Profile in Appendix I; contemplate how your answers affect content, structure, and tone.
❖ Understand your and your receivers' specific purpose(s); only then can you influence receivers to react in the way(s) that you want and avoid extraneous excursions to unrelated topics that destroy concision and clarity, and that label you as a bad writer and thinker.

- Understand the subject enough to enable you to start writing, and start early to allow time for your subconscious smarts to do the hard work and to revise.

- **Write backwards, from the end of your document to the beginning, from the details to the abstract**

- Start with the details: an appendix, e.g., in a larger document, the body in a smaller document. You will learn more about the subject, allowing you to ...
- Compose an insightful, enticing, and reflective opening sentence/paragraph/summary more quickly. Then ...
- Write a title, subject line, and/or table of contents—which can be a summary in the form of an outline-- as required by the length and complexity of the document.
- Apply, diligently, and without compromise, the techniques of senders that follow. Only then will your document be clear, concise, and purposeful.

> Pete 6/13/13 1:38 PM
> **Comment:** Exactly the opposite of what you learned in grammar school. Trust me; backwards is really bestwards. (It's OK to make up new words if their meaning is obvious.)

- **Think after you write**

- Check for quality by completing the one-word outline to uncover disunities, illogical arrangements, and opportunities to chunk; and by reading aloud to uncover and correct clumsy syntax.
- Revise, revise, revise ...

> Pete 6/13/13 1:38 PM
> **Comment:** See CHAPTER V for more.

> Pete 6/13/13 1:38 PM
> **Comment:** Reading aloud brings another of your senses—hearing—into play for your benefit.

> Pete 7/23/13 5:40 AM
> **Comment:** It bears repeating: I don't know anyone whose first draft is worth reading, much less perfect.

- **Think after receivers react**

- Decide if your document influenced receivers in the ways you intended when you completed the Receivers' Profile, or didn't. Regardless, think about what you did right and wrong, and reinforce the right and correct the wrong in subsequent documents. Too many of us focus on what we did wrong rather than right, yet right is at least as important.

Academicians think the writing process is:

Who says what→to whom→in what channel(s)→for what effect.

The process in the real world is reversed:

For what effect (your and receivers' purposes)→on whom (your receivers) →do you say what

(content)→in what order (structure) →using what words (tone)→and test (did it work as planned, and, if not, adjust subsequent sendings)

> Pete 6/13/13 1:38 PM
> **Comment:** *Effect* is usually a noun meaning *result;* it is sometimes a verb meaning *to cause to happen.* *Affect* is almost always a verb meaning *to influence or produce an effect. Affect* can be a noun meaning a feeling.

Many of my students resonate with my ten-minute drill:

FIRST:

1. Complete the Receivers' Profile in writing; one- or two-word answers are sufficient.
2. Contemplate how and why your answers will impact content, structure, and tone.
3. Fix on the key points...select the one or two that are most important to you and the reader. Start your piece with those points.
4. Order the other points by importance.

Pete 6/13/13 1:38 PM
Comment: Again, think before you write; you will create a better document in less time—quality and productivity in one easy habit.

THEN (at 11 minutes):

5. Write a quick first draft without considering grammar or syntax; put your thoughts on paper in the order you've selected. Start writing with the details, work toward the abstract.
6. Let your subconscious smarts kick in; put the piece aside for hours, days if possible.
7. Rewrite often by referring to the Receivers' Profile (don't fall in love with your words).
8. Consider your writing complete when it is true to the profile.

7. ***Buy and use at least two reference books:*** a recent dictionary/thesaurus and a concise handbook for writers, and keep them handy. Or, **tap into your computer's dictionary/thesaurus** and its explanation of highlighted words and phrases; it's a great way to expand your vocabulary—the root of better writing and thinking-- and knowledge of grammar and syntax.

Pete 6/13/13 1:38 PM
Comment: See APPENDIX VI, FURTHER READING for specific suggestions.

Pete 6/13/13 1:38 PM
Comment: Grammar and syntax are often thought of as synonyms, but aren't. Grammar is the rules of the language's inflections or other means that show the relation between words; syntax is the grammatical arrangement of words to show their relationships and connections, logically, we hope.

8. ***Find your most congenial/comfortable/productive time and place.*** You can't compose a proper sending if you're distracted for any reason: noise, hunger, caffeine fit, etc.

Pete 6/13/13 1:38 PM
Comment: See APPENDIX V for useful abbreviations.

CHAPTER III: EMBRACE THE SEVEN HABITS OF SUCCESSFUL RECEIVERS

Garbage in, garbage out applies equally to your mind and computer. Ergo, listening to others and yourself (called studying) is essential to learning the subject and to creating sendings that do their jobs. Some tips:

1. **Compartmentalize your mind by clearing it of clutter**; your other thoughts can wait.
2. **Listen actively and critically**; pay attention to what you're reading (aka studying, or listening closely to yourself) or hearing (aka active listening to others).
3. **Organize as you listen;** bring details up to a larger abstraction.
4. **Rephrase periodically as you listen**; you'll tell the sender that you're paying attention and you want to understand the message.
5. **Take notes;** you don't need to jot down each small idea, but you do need to record the bigger abstractions, forcing you to interpret what the sender is conveying.
6. **Enable speakers to finish their sentences without interruptions;** doing so proves that you are paying attention and the sender is important, while interrupting proves that you are thinking more about how you will react or rebut and you don't care about the sender. Put yourself into the senders' minds and decide how you'd like to be treated, then do unto others ...
7. **Be positive;** tell senders when their sending meets the criteria for good; also tell senders when the sending doesn't by suggesting ways that it can.

Pete 6/13/13 1:38 PM
Comment: Latin for therefore, for that reason, consequently, made popular by Renee Descartes, the French philosopher.

Pete 6/13/13 1:38 PM
Comment: I know I repeated this idea; it's worth it.

Pete 6/13/13 1:38 PM
Comment: ... is an ellipse and indicates that I've omitted something.

CHAPTER IV: INTERNALIZE THE SEVENTEEN TECHNIQUES FOR SUCCESSFUL SENDINGS

1. Create clear, concise sentences of varying lengths and types.

A sentence is a group of words containing a subject (person or thing being talked about: a noun or pronoun) and a predicate (what is said about the subject: a verb and its modifiers).

Most senders use only three types of sentences: simple, containing a single complete thought, aka an independent clause (We appreciate your prompt reply to our proposal.); compound, containing two or more independent clauses (We like the design, but need more options.); and complex, containing at least one independent clause and one dependent clause—a group of words that is not a complete thought and is not a complete sentence (You may collect the refund if you apply within ten days.)

Pete 6/13/13 1:38 PM
Comment: The subject, we, is understood here. I could have written: We like the design, but we need more options.

Pete 6/13/13 1:38 PM
Comment: A complete thought.

Pete 6/13/13 1:38 PM
Comment: Not a complete thought.

You can start a sentence with "and" or "but", and you can end a sentence with a preposition, despite what you learned from misguided teachers and professors.

Keep the average sentence length to 20-25 words or fewer, but vary the length to avoid boring receivers.

Avoid starting a sentence with *There: There is sensitive equipment for this procedure.*
This procedure requires sensitive equipment.
Sensitive equipment is available for this procedure.

Pete 6/13/13 1:38 PM
Comment: Did you notice that I changed the meaning?

Avoid starting a sentence with *It: It was his bad attitude that got him fired.*
He was fired because of his bad attitude.
His bad attitude caused him to be fired.
His attitude was so abrasive that he was fired, and good riddance.

Pete 6/13/13 1:38 PM
Comment: Passive voice, always. Be active when you can.

Pete 6/13/13 1:38 PM
Comment: How bland.

Pete 6/13/13 1:38 PM
Comment: I changed the tone, added attitude.

Delete needless prefaces: *I am writing this letter because I am applying for the job of Senior Accountant.*
I wish to apply for the job as Senior Accountant.

Pete 6/13/13 1:38 PM
Comment: The receiver can see that you wrote a letter; you don't need to say so.

Cut unnecessary words by revising the sentence's structure: *As far as artificial intelligence is concerned, the technology is only in its infancy.*
Artificial intelligence is in its infancy.

2. Develop topic sentences for paragraphs that express attitude and are supported by examples and details.

Paragraphs are separate blocks of prose that typically are made up of a topic sentence (the controlling idea that keeps senders on point); support for the topic sentence (examples, explanations, etc); transitional words or phrases; and a conclusion or transition to the next paragraph. Another way to depict the structure of a paragraph: A (the abstract idea that is the topic sentence), B (the body or development of the abstract idea), and C (the closing or conclusion). Still another is The Preachers' Maxim: Tell 'em what you're planning to tell 'em, tell 'em, and tell 'em what you told 'em.

Topic sentences that do not/do express an opinion or attitude:

> *Our firm employs 250 people. (So what?)*
> Working within a firm with 250 employees opens opportunities for technical, managerial, and social interactions that I never thought possible. (Aha, now I get it.) Why just the other day my boss asked me to ...
>
> *Writing is a complex skill. (No kidding.)*
> I never realized that writing was so demanding of my intelligence and time. (A specific epiphany.) For example . . .
>
> *A Lexus is a great all-around car. (Yawn)*
> A Lexus ES is safe, comfortable, and reliable . . . the perfect car for me. (OK, now I get your point.) It's safe ...(etc). It's comfortable . . . (etc), and it's reliable . . . (proof).

A manufacturer of pollution control equipment opens a sales letter with this paragraph:

Please find enclosed a brochure detailing the air abatement equipment supplied by XYZ Technology System. Our experienced staff of experienced professional engineers custom designs cost effective emission control system for each application. Our list of supplied equipment include: packed absorption towers, tray scrubbers, activated carbon adsorption towers, cyclones, demisters and System R our featured patented Venturi scrubber. Its advantages include:

(better) Perhaps we can help you in your search for cost-effective air pollution control equipment.
Our absorption towers, tray scrubbers, cyclones, and other equipment—all described in the attached brochure—are custom-designed by experienced engineers and manufactured in our modern factory to minimize costs and maximize performance. You may be particularly interested in System R, a patented wet scrubber. Its advantages include:

Pete 6/13/13 1:38 PM
Comment: Notice how the three points in the topic sentence are supported in the body in exactly the same order. We call that cohesion, a form of parallelism. See below.

Pete 6/13/13 1:38 PM
Comment: Ho hum; where's the attitude?

Pete 6/13/13 1:38 PM
Comment: How many *experienced* are needed?

Pete 6/13/13 1:38 PM
Comment: An obvious disunity; the second sentence is unrelated to the topic sentence.

Pete 6/13/13 1:38 PM
Comment: List is singular; the verb plural.

Pete 6/13/13 1:38 PM
Comment: Needs a comma.

Pete 6/13/13 1:38 PM
Comment: A mild attitude. To be more aggressive: I'm certain that we can help you find cost-effective ...

Pete 6/13/13 1:38 PM
Comment: Passive with reason?

(In active voice) Our experienced professional engineers custom-design a wide range of extremely effective and efficient abatement equipment—towers, cyclones, and more—all described in the attached brochure. You might be particularly interested in System R patented Venturi wet scrubbers. Its advantages include:

Which version do you prefer?

3. Be parallel whenever you can.

Parallel sentences and other parts of a document show similarities of thoughts and ideas by using similar grammatical constructions, adding to clarity and concision.

(Not parallel*) She began walking back and forth and to frown worriedly.*
(Parallel) She began to walk back and forth and to frown worriedly.

(No*) The new employee is enthusiastic, skilled, and you can depend on her.*
(Yes) The new employee is enthusiastic, skilled, and dependable.

(No) *We serve breakfast, lunch, and catering.*
(Yes) We serve breakfast and lunch, and will cater your event at your home or office.
(Yes) We serve breakfast and lunch here or at your home or office.

Parallelism extends to data:
Last season the Cowboys were picked to finish 11th in the big 12, but finished11 and 2. (Where did they
rank in the Big 12?)Or . . . finished with 11 wins and two losses.

Last season the Cowboys were picked to finish 11th in the Big 12, but finished second.

Parallelism extends its reach to chapter headings, headlines and subheads in the text, and the first words in lists. Did I do that in this book?

4. Prefer active voice, but revert to passive when you must

Active voice: the subject of the sentence is acting upon something; passive voice: the subject is being acted upon.

(Active) John is filing the letter.
(Passive) The letter is being filed by John.

Active is clearer--the actor is identified –and more concise—one fifth to one-third fewer words are typically needed to express the thought—

Pete 6/13/13 1:38 PM
Comment: I added attitude. OK?

Pete 6/13/13 1:38 PM
Comment: Two verb forms.

Pete 6/13/13 1:38 PM
Comment: One verb form

Pete 6/13/13 1:38 PM
Comment: Two adjectives and one independent phrase. (A complete thought, aka a. sentence.)

Pete 6/13/13 1:38 PM
Comment: Three adjectives.

Pete 6/13/13 1:38 PM
Comment: Three nouns, but two are things (meals) and one is a service (catering). Can you serve catering?

Pete 6/13/13 1:38 PM
Comment: Numbers ten and above are typically spelled out; nine and below not. Some publications and firms reverse that rule, and some publications, notably *The New Yorker,* spell out all numbers. Be consistent within your document.

Pete 6/13/13 1:38 PM
Comment: John is the subject and we know what he's doing.

Pete 6/13/13 1:38 PM
Comment: The letter is the subject and we know what's happening to it.

Pete 6/13/13 1:38 PM
Comment: Not 'less words'. Use *fewer* with nouns that can be counted, *less* when they can't be.

and more dynamic with its sense of energy and motion. (I used too many dashes, didn't I?) Be active when you can and passive when you must, e.g.[1] when you don't know or want to subdue the actor, or when the actor is less important than the item being acted upon, and when you want to deflect responsibility. *The mistake in the document was made by Jones,* instead of *Jones made the mistake in the document.*

(Active) The golf ball hit the senator.
(Passive) The senator was hit by the golf ball.

What's more important, the golf ball or the senator?

Many senders mistakenly feel that the passive voice denotes authority and intelligence. In fact, it denotes pompous evasion and muddiness, and it confuses and bores receivers. An example:

It has been decided that, effective immediately, the doors to the school will be locked at 9PM each day. Only those employees who have been issued a special pass will be permitted to enter thereafter. All others will not be admitted, and will be docked a day's pay. (48 words)

From this memo, it cannot be determined (note the passive voice; inevitable when starting a sentence with *it.* I could have written *From this memo, receivers cannot determine ...*). Receivers cannot know by whom the action was decided, for what reason the action was undertaken, by whom the doors will be locked, by whom or under what circumstances a special pass will be issued, or by whom the employee's pay will be withheld. (Note that I asked all these questions in passive voice). I'll rewrite the memo in active voice:

The board decided that, effective immediately, Joe Jones will lock the doors to the school at 9PM each day, Sam Smith will issue special passes to those who need them, and Katie Killjoy will dock offenders a day's pay. (39 words, 9 words—19 percent--fewer than the passive paragraph.)

5. Convert lazy nouns and adjectives into active verbs.

Nouns and adjectives that can be verbs are called "smothered"; they stifle the smooth progress of the document by adding

---

[1] *Exempli gratia*—Latin meaning *for example*. Please don't confuse it with i.e.--*id est*—Latin for *that is to say*. See APENDIX V

another verb, the lazy verb, and several unneeded words by throwing the sentence into passive voice:

Authorization for the trip was given by Joe. (Authorization, noun, can be a verb.)
Joe authorized the trip.

My new assistant is negligent in the details of her design. (Negligent, adjective, can be a verb.)
My new assistant neglects the details of her design.

The following nouns and adjectives, among others, can be converted to verbs, adding clarity, concision, and action to sendings:

Administration, administer
Advancement, advance
Authorization, authorize
Performance, perform
Concession, concede
Confrontation, confront
Determination, determine
Illustration, illustrate
Implementation, implement
Management, manage
Negotiation, negotiate
Quotation, quote
Realization, realize
Relation, relate
Transmittal transmit
Utilization, utilize

Examples, with the **lazy verb:**

Nancy ***provided*** *her authorization for the purchase.*
Nancy authorized the purchase.

Sam ***was*** *responsible for implementation of the program.*
Sam implemented the program.

Pete 6/13/13 1:38 PM
Comment: Did I change the meaning?

Pete ***came*** *to the realization that the job was perfect for him.*
Pete realized that the job was perfect for him.

Joe ***made*** *an attempt to answer the question.*

Joe attempted to answer the question.

He **provided** a solution to the problem.
He solved the problem.

The thief **issued** an apology.
The thief apologized.

ABC will **supply** the equipment needed by the troops.
ABC will equip the troops.

6. Always be cohesive, unified, and apply a related technique ...

7. Use lists when appropriate, but not so much that your document looks like an outline.

Cohesion—the logical sequence of thoughts-- and unity—the completion of a thought before moving on to the next-- are your most important assets. Yet senders treat them with shameless disdain by wandering aimlessly to irrelevant thoughts, another way of saying that they go off-point; they lose sight of their purpose.

Pete 6/13/13 1:38 PM
Comment: Note the passive voice for a reason. I could have written: Your most important assets are ...The assets are more important than you are. Do you agree?

Bullets are details that support an abstraction, just as the body of a paragraph supports the topic sentence, and are typically arranged by importance from top to bottom; numbers typically display a sequence of events such as the steps of a process or an event. Either type can be arranged vertically—more common—or horizontally. Caveat: Limit bullets in any one list to three or four; receivers won't understand or remember longer lists (your telephone and social security numbers are in groups of three or four for this reason). Your numbered list can be longer, but not so long that receivers lose their way and become confused.

Pete 6/13/13 1:38 PM
Comment: Note the parallelism.

An incoherent paragraph from an agribusiness that benefits from bullets:

Chemical pesticides are both ineffective and hazardous. Because none of these chemicals has permanent effects, pest populations invariably recover and require re-spraying. Repeated applications cause pests to develop immunity to the chemicals. Furthermore, most of these products attack species other than the intended pest, killing its natural predators, and thereby actually increasing the pest population. Above all, chemical residues survive in the environment (and living tissue) for years, often carried hundreds of miles by wind and water. And these chemicals are expensive. At a time when our company seeks to reduce expenses, we should look for more economical solutions.

Pete 6/13/13 1:38 PM
Comment: I like the attitude.

Pete 6/13/13 1:38 PM
Comment: Not mentioned in the topic sentence.

99 words

The paragraph isn't cohesive: Receivers expect "ineffective and hazardous" to be supported in that order, but they aren't. The body sentences address cost, ineffective, ineffective, hazardous, cost, conclusion--- a stew of ideas, aka a disunity that can be cured by chunking.

Pete 6/13/13 1:38 PM
Comment: Chunking is writer-speak for placing similar thoughts together, i.e. not spreading them or revisiting them in the document except for emphasis.

A better way:

Chemical pesticides are ineffective, hazardous, and costly. Their effects are short-lived and pest populations invariably recover quickly from a single spraying, requiring additional applications. As a result, pests soon become immune. Furthermore, most pesticides kill species other than the intended pests, killing natural predators and actually increasing the pest population. Above all, chemical residues survive for years in the environment and in living tissue, and are often carried hundreds of miles from the farm to contaminate a large area. Finally, pesticides are expensive, just what we don't want at this time. We need to search for more effective, less hazardous, and more economical alternatives.

Pete 6/13/13 1:38 PM
Comment: Note the transitions: *as a result, furthermore, above all, and finally*. They are important to lead the reader into the next thought. Use them, and recognize that they signal the need for a list.

104 words, but clearer by being more cohesive and unified.

The same thoughts in vertical bullets:

Chemical pesticides aren't the solution to our pest problem. They:

Pete 6/13/13 1:38 PM
Comment: Where's the attitude? Try: Chemical pesticides can't solve our worsening problem with pests. And note that the bullets answer *why* and *how*.

- ❖ Don't perform for long: Pests invariably recover in only a few weeks from a single spraying, and when they do they can be immune, requiring additional sprayings of different chemicals. And pesticides can attack natural predators, actually increasing the pest population, requiring other measures such as more spraying with stronger chemicals.
- ❖ Are hazardous to the environment and living organisms: Their residues survive for years, and can be carried for hundreds of miles to other areas, greatly increasing our liabilities.
- ❖ Cost way too much for their supposed benefits: In short, they are a poor value and flunk any cost/benefit analysis.

The bottom line is simply that it behooves us to search for more acceptable alternatives, two of which are ...

Pete 6/13/13 1:38 PM
Comment: Complete the thought for the receiver.

109 words. If you were a CEO reading this, which version would you prefer?

Pete 6/13/13 1:38 PM
Comment: Not a sentence, obviously, but Ok to meet my purpose to point out the number of words.

Here's a perfect example of too many bullets that can benefit easily and quickly by chunking and bringing smaller ideas into larger concepts:

We serve the following industries:

- *Alternative fuels*
- *Automotive*
- *Chemical and petrochemical*
- *Consumer products*
- *Food and beverage*
- *Gasification*
- *Industrial infrastructure*
- *Manufacturing*
- *Midstream and pipelines*
- *Oil and gas*
- *Pharmaceutical*
- *Refining*

OK, now that you've plowed through twelve disjointed bullets, what industries are served by this giant of engineering? Can I boil twelve bullets to four or fewer? Are the industries cited in fact industries?

Pete 11/18/13 1:04 PM
Comment: Not *less,* which is used before *mass quantities; fewer is* reserved for items that can be counted. I've repeated this rule because it is one of my pet peeves. Other writers tell me I am picking nits.

We serve the following industries:

- Energy: oil and gas, refining, alternative fuels, coal gasification, midstream and pipelines;
- Manufacturing: pharmaceuticals, chemicals/petrochemicals, consumer products, foods/beverages, automobiles; and
- Industrial infrastructure (what does that mean? Surely it is not an industry.)

Which version gives you a better understanding of what this firm offers?

1. **Punctuate properly; it's needed for clarity and concision**

Here are simple rules that apply to almost every document:

COMMAS are placed:

a) Between all words in a series, including the last two, unless the last two are inseparable pairs such as "ham and eggs" or "love and marriage". Meanings can be mangled if they aren't:

Write name, address, age, sex and housing requirements. (sex requirements?)

It appears that ABC Company knew its right to occupy the land was precarious, far from sound and tenuous. (far from tenuous?)

The money is to be divided among Jack, Susan, Paul and Jose. (Paul and Jose are cheated.)

b) Before conjunctions such as *and, but, or, nor,* and *so* when joining independent clauses (complete thoughts, i.e. sentences without periods). Use a semi-colon when connecting independent clauses without inserting a conjunction.

Our customer knew very little about our services, so he asked for a brochure.

Joe worked hard to negotiate the contract, and he was rewarded with a paid vacation.

c) Around parenthetical insertions; do not place a comma at one end and not the other.

Our new design process, which is based on teamwork, has worked well.

Joe, who has succeeded in other areas, is out of his element here.

All of us, to tell the truth, were amazed how well he could putt.

Our second report, on the other hand, satisfied the customer.

d) After most introductory word(s) and phrases.

In conclusion, we feel that our proposal answers your needs completely and efficiently.

Indeed, we agree totally with your position.

e) Within items in a series that are separated by semi colons (the internal comma).

The company's assets include $20 million in land, buildings, and equipment; $44 million in cash, stocks, and bonds; and $100 million in receivables and inventory.

Our proposal covers all design, construction, and shipping; labor, material, and tooling; and travel expenses for foremen and superintendents.

SEMI-COLONS are used much like commas (to separate elements in a series), periods (to mark the end of an independent clause), and colons (to indicate that the remaining parts of the sentence are closely related). Place a semi-colon:

a) To separate related independent clauses joined without a coordinating conjunction such as *and* or *but*:

The engineer completed the design; his boss was pleased.

The engineer completed the design and his boss was pleased.
The engineer completed the design, and his boss was pleased.

b) To separate elements in a series when the elements themselves contain internal commas:

The engineer completed the design, wrote the specs, and visited the site; she then played golf, dined, and drove home.

Our sales engineer plans to visit customers in Pittsburgh, PA; Columbus, OH; and Portland, OR.

c) To separate extraordinarily long elements in a series even if they don't have internal commas:

Our engineers recommend that the breakers be installed on an extra-thick concrete slab to protect against the possibility of an earthquake; that the entire installation be surrounded by a ten-foot high chain link fence topped by barbed wire to protect against vandalism; and that the dielectric gas be exchanged every ten years to assure reliable operations and protect against unwanted grounding.

Pete 6/13/13 1:38 PM
Comment: A candidate for a list.

d) To connect two clauses when the second begins with a conjunction such as *however, indeed, thus;* or by connecting transitional phrases such as *as a result, in that case, on the other hand.*

The candidate for the job is a fine designer; however, he cannot write clearly and concisely.

Without question, the candidate for the job can design LEED buildings; on the other hand, he seems unable to express himself in writing or speaking, a serious drawback.

Our project engineer wrote a misleading proposal; as a result, we were not considered for the contract.

Pete 6/13/13 1:38 PM
Comment: Note the three transitions in the three examples: *however, on the other hand*, and *as a result*.

COLONS are typically marks of introduction that alert readers that additional, related, explanatory, amplifying information is on the way. Place a colon:

a) To introduce a clause or phrase that "pays off"—completes-- the thought of the clause or phrase to its left, often with a list:

Our geologist visited the proposed site and found: fossil remains of fish from an early era, limestone sinkholes, and a small stream that was unmapped.

Pete's writing classes come down to this: clear and logical expression, efficient use of words, and a sharp focus on defined purposes.

b) To emphasize a point:

Our proposed building is designed around one benefit: energy efficiency.
My boss demands only one thing from me: perfection.

c) To separate independent clauses, replacing a semicolon.

We offer the finest breakers available: They are reliable and cost less to own and operate.
Nobody questions the competence of our environmental scientists: We hire only graduates from the most prestigious universities.

A note on punctuation: Capitalize the initial word of the second independent clause, but lower case the initial word when the colon is followed by a dependent clause or list.

Our CEO is paid handsomely: He earns more than $2.4 million per year. (Two independent clauses)

Our CEO is paid well: more than $2.4 million per year. (One independent and one dependent clause)

Our CEO is paid in three ways: salary, bonuses, and stock options. (A list)

DASHES are so versatile that they can replace commas, semicolons, colons, and parentheses whenever and wherever—almost--the writer wants. Place a dash:

a) To indicate an abrupt change or break in thought:

Our sales engineer was happy with the proposal—the customer, however, felt differently.
My boss is a great engineer—but he can't write a simple sentence, much less a paragraph.

b) To replace commas or parentheses to attract attention to subordinate or amplifying material:

The project engineer focused on these four steps—study, design, manufacture, and closure--needed to finish the project on time.

I can't tell you enough how much I enjoyed your presentation—its usefulness, the visuals, and your enthusiasm—on the benefits of LEED design.

c) To replace colons to introduce a list or defining phrase:

Our company is an innovator in at least three important areas--safety, reliability, and efficiency.
We want to gain market share—to at least 25%.

APOSTROPHES are used, and misused regularly, to form most possessives and contractions, and to indicate omissions.

a) Possessives:

--If the possessor is singular: The firm's design, marketing's role, the estimator's guess, the building's façade.

--If the possessor is plural: The firms' design (or designs), marketings' and sales' (sale's is incorrect) role (s), the buildings' façade (s) ...

A double blip by a local restaurant: *FATHER'S DAY RESERVATIONS! Dad's will receive a $10 certificate from XXX.* (Only one Father? Why is Dad possessive?)

Another common blip: MENS ROOM or MENS DEPARTMENT. You can't make a plural out of a plural.

b) Contractions:
--wasn't, isn't, they're, she'd, who's, ass'n, dep't. Pete's (as in Pete is ... or Pete has ...or it could be possessive),

c) Omissions in numbers
--back in the '90s, the class of '04 ...

A headline: *DEAD SONS PHOTOS MAY BE RELEASED* (taken freely from *Eats, Shoots and Leaves, by Lynne Truss*)

Pete 6/13/13 1:38 PM
Comment: Photos of dead sons?

.

HYPHENS are used most often to connect the parts of compound nouns and modifiers:

Cost-effective (as in *cost-effective design*, not *the design is cost effective*)
Spin-off
Eye-opener

On-ramp
Cross-reference
Double-check
Designer-foreman

Standard practice for hyphens varies widely—another example of stylistic preference. Nevertheless, the sender is responsible for creating clear meanings, a particularly vexing challenge with compound adjectives. Rule: If one element of the compound doesn't make sense without the other, add the hyphen. See cost-effective and double-check, above.

Caution: Many compound nouns have lost their hyphens and, because of common usage, are one word: football, workplace, cooperate—and others are two words—fuel oil, health care, pre-eminence. Refer to a recent handbook and dictionary for current usage, and consider these guidelines:

--Decide which version—two words, hyphenated, or one word—will be clearest for receivers.
--Note how other good writers in respected journals use hyphens, and follow their lead.
--Unite words that go together naturally; you can fearlessly pioneer changes in the language.

NUMBERS from one to nine are typically spelled one to nine, expressed in numerals for 10 and above. However, the protocol for technical journals is to use numerals exclusively, and the protocol for other publications is to spell out all numbers. Other rules:

--Be consistent throughout the document.
--Spell out a number that starts a sentence.
--Use numerals for dollar amounts.

2. Use heads and subheads that benefit senders and receivers

Business documents are read like textbooks, not novels; readers jump back and forth to sections that interest them most at the time. For example, a CEO reviewing a proposal may jump to PRICE, the Chief Engineer to TECHNICAL APPROACH, and so on. Heads and subheads enable them to do so easily and quickly. Other benefits include:

√ Receivers are alerted to a transition; they'll be alerted that one topic is ending and another is beginning. They can decide if a section is worth reading, saving time and raising productivity; they understand how the document is organized, creating a sense of order; they are less intimidated by a document that is broken into sections; and they are more likely to remember important points, especially if the heads are editorial. (Note how I used semis to separate items in a string.)

√ Senders are kept on course and avoid pesky transitions in the text such as: In addition, Another point worth making, and the like. Your heads and subheads are the transitions.

THE THREE TYPES OF HEADS BASED ON CONTENT

1. Instructional, e.g., Introduction, Conclusion, Background, Price (as I did in the initial paragraph in this section) Many such headings are of limited use, of marginal value above no heading at all.
2. Informational, e.g. How to Format Your Document instead of Document Formatting; or, Scrubbers Would Reduce Emissions of Sulfur Dioxide instead of The Role of Scrubbers; or, The Effects of Acid Rain on Lake Trout instead of Acid Rain.
3. Editorial, e.g., We Had a Great Year instead of Earnings; or, Our Air is Cleaner than Ever! instead of Clean Air is Important; or, Acid Rain is Killing Millions of Trout in our Lakes instead of The Effects of Acid Rain on Lake Trout. [Is the meaning changed?]

Pete 6/13/13 1:38 PM
Comment: The head above is instructional and is used to point the way for the reader.

THE TWO TYPES OF HEADS BASED ON FORMAT

- **Floating**, placed above the text, as demonstrated above; and
- **Run-in** as part of a sentence, as demonstrated here.

Points to Ponder

√ Be parallel.

√ Be as specific and concise as possible [remember our definition of concise].

√ Avoid starting the sentence after the heading with "this", "it" or some other pronoun that refers to the heading. Instead, make the sentence's meaning independent of the head.

3. Don't be strung out on strings, nouns and others

Nouns, and, to a lesser extent adjectives and adverbs, are too often strung together so that the first words modify the latter. However, strings—especially if they are more than three words long--are difficult to read and understand, eroding clarity. Avoid them by substituting descriptive [usually prepositional] phrases that clarify the relationship between the thing being modified--the main or essential noun--and the modifier—a word that qualifies or changes in some way the sense of another word. Here are several examples:

--*Be sure to leave enough time for a training session participant evaluation.* [An evaluation of or by the participants? We don't know.]
--Be sure to leave enough time for participants to evaluate the training session. Or: Be sure to leave enough time to evaluate participants in the training session.

(Title of a book)
--*The Great Wine Book.* (Is it a great book about wine, or a book about great wine? Could the ambiguity be purposeful to sell more books?)

[Title of a report on wetlands]
--*Corps Descriptive Method Functional Method Assessment Considerations/Rationale.* (I have no idea what the writer means. Do you?)

--*The company specified PPG's Teslin synthetic printing sheet.*
--The company specified Teslin printing sheet by PPG.

--*The president of Edina, Minnesota-based Markham Capital Management thinks confused small investors present a lucrative market.*
--The president of Markham Capital Management, which is based in Edina, Minnesota, thinks confused small investors present a lucrative market.
--The president of Markham Capital Management thinks small investors, many of whom he describes as confused, are a lucrative market. The firm is based in Edina, MN.

--*XYZ is a privately held, 60-person environmental engineering firm located in ABC PA which for 28 years has provided planning, site development, design, permitting, operational and construction related engineering services to its 200+ private and public western Pennsylvania client base.*
--XYZ 's 60 employees provide planning, site development, design, permitting, and operational/construction-related services to 220+ public and private clients in western PA. The firm is headquartered in ABC PA and has been in business for 28 years.

--*Babe Ruth Birthplace Museum executive director Mike Gibbons displays a rare 1914 baseball card at the museum in Baltimore.*
--Mike Gibbons, executive director of the Babe Ruth Birthplace Museum in Baltimore, displays a rare baseball card printed in 1914.

11. Use footnotes and endnotes to help senders and receivers

Your narrative proceeds full tilt and on track, without diversions to other topics, when footnotes and endnotes offer curious readers extra, sidebar information that they may want; they also allow less curious readers to ignore that same info. Everybody wins.

Pete 6/13/13 1:38 PM
Comment: Think cohesion.

Pete 6/13/13 1:38 PM
Comment: *They* is a pronoun that I want to refer to endnotes and footnotes; does it?

Footnotes are located on the bottom of the page on which they are cited, and endnotes are bunched together at the end of the chapter or document.[2] Each can be:

Pete 6/13/13 1:38 PM
Comment: Readers are faced with two options, each cumbersome and intrusive: drop to the bottom of the page or leaf to the end of the chapter or document. I'm open to suggestions for good alternatives.

- Explanatory, typically containing incidental intelligence not considered important enough to be part of the narrative.
- Editorial, to express your opinion.
- Reference, aka bibliographical, typically citing other sections of the document or other publications that offer supplemental information.

Footnotes and endnotes can be useful in at least one other way that may not be so obvious. A document sprinkled liberally with *ibids* and *op. cits*[3] emits an aura of (false) scholarship that impresses some (gullible) readers, adding (dubious) credibility to writers. Academics love this ridiculous and pompous ploy, but it has no place whatsoever in business.

Pete 6/13/13 1:38 PM
Comment: Note my similar comment about passive voice.

Suggestions re format: Separate footnotes from the main text by several spaces; number the footnotes and/or endnotes consecutively throughout the section or document; and highlight the key in the text as much as your word processing software will allow (readers dislike hunting for the key).

12. Avoid run-on and choppy sentences.

---

[2] Many writers consider in-text citations to be a form of footnote. I don't; they interrupt the narrative as a parenthetical note, diverting the reader from the main ideas. Avoid them.

[3] You probably forgot the meanings of these Latin terms long ago, as I have, and I could avoid telling you and force those who are interested to look them up in a dictionary. But by doing so I would violate my rule to ease the readers' life. So: ibid means "in the same work", and is used instead of the complete reference; op cit means "in the place cited", and is used with the name of the author but not the page number. Clear as proverbial mud, right? Avoid these terms if at all possible; your readers likely won't know their meanings either.

Run-on sentences cram too many ideas together without needed breaks or pauses between thoughts:

The key factor in the plummeting injury rate lies in the steel industry's underlying belief that safety must be regarded as an integral element of operations because instead of just producing steel and adding safety as an afterthought, the companies plan the safe production of steel, and everything from engineering through education and enforcement is designed to that end.

The plummeting injury rate did not just happen. Perhaps the key factor lies in the steel industry's underlying belief that safety must be regarded as an integral element of operations. Companies don't just produce steel and think of safety as an afterthought; instead, they plan the safe production of steel, and everything from engineering through education and enforcement is designed to that end.

Pete 6/13/13 1:38 PM
Comment: Note the semi colon followed by a transition and comma, a useful construction.

Choppies are short sentences that remind receivers of their experiences reading Dick and Jane stories in third grade; they are tedious and label senders as immature:

The world's forests are now disappearing. The rate of disappearance is 18 to 20,000,000 hectares a year (an area half the size of California). Most of this loss occurs in humid tropical forests. These forests are in Asia, Africa, and South America.(Is it 18 or 18,000,ooo?)

The world's forests are now disappearing at the rate of 18 to 20,000,000 hectares a year, an area half the size of California. Most of the losses are occurring in the humid tropical forests of Asia, Africa, and South America.

Rule #1 to change choppies to acceptable sentences: Sentences that start with this, that, these, and those—or with the same words or thoughts that ended the previous sentence—often can be combined into one sentence.

(Bad) *The Polyad process employs a unique adsorbent called Bonopore. Bonopore is a macroporous polymer which is superior to other existing adsorbents. (21 words)*

(Better) The Polyad process employs a unique and superior adsorbent, Bonopore, a macroporous polymer. (13 words)

(Better?) The heart of the Polyad process, Bonopore, is a unique macroporous polymer. (12 words)

(Another way?) Bonopore, a unique macroporous polymer, is the heart of the Polyad process. (12 words)

Pete 6/13/13 1:38 PM
Comment: Is *Bonopore* more important than the process?

Rule #2. Vary the lengths of sentences to avoid boring readers. Many writers feel that a good average length is 20-- 23 words, but avoid writing for an average. Writers in The New Yorker, for example, all very adept and experienced, often create sentences of 150 words or more and use colons, semi colons, and dashes to create pauses and to closely connect thoughts. They also write sentences of only two or three words—a subject and predicate—for variety: I am, we are, I am going, etc. I often write 'sentences' of one word—Maybe, Never, Baloney, are good examples--when I want to emphasize a thought, and I set the word off in a separate paragraph. I am, of course, ignoring the rule that a sentence is a subject and a predicate, but I can get away with that; maybe you can't.

Pete 7/23/13 8:42 AM
Comment: One word cannot be a sentence, so I added scare quotes to show that I know that.

The choppiness of the ensuing paragraph can be converted to readable by connecting thoughts and adding **transitions**.

Pete 6/13/13 1:38 PM
Comment: I used passive voice but could easily use active: You can convert the ensuing paragraph from choppy ...

Writing is a difficult and important skill to master. It requires long hours of work and concentration. This time and effort are well spent. Writing is indispensable for success. Good writers derive great pride and satisfaction from their effort. A highly disciplined writing course should be a part of every student's curriculum.

(52 words)

Writing is a difficult but important skill that requires long hours of work and concentration to master. **However**, this time and effort are well spent because writing is an indispensable tool for success. **Moreover**, good writers derive great pride and satisfaction from their efforts. **Therefore**, a highly disciplined writing course should be taken by every student.

Pete 6/13/13 1:38 PM
Comment: Passive. Can you make it active? Is this better? Writing is a difficult and important skill that anyone can master by expending long hours of work and concentration. Still passive, isn't it. Try: I have learned that writing is ...

Pete 6/13/13 1:38 PM
Comment: To be active: Every student should take ...

(56 words)

To express the same thoughts in active voice: I have learned that writing well requires long hours of concentrated work; I've also learned that the time and effort are well spent and are needed to be successful. Therefore, I have enrolled in a disciplined writing course.

(38 words)

13. Write abstracts and summaries last

The differences between abstracts and summaries: An abstract is generally shorter and either informative or descriptive; a summary is longer and can be informative, descriptive, or editorial. Abstracts and summaries cover the main points in the document, and generally include statements of purpose and the most important points needed by decision-makers. For your purposes, the differences are a matter of semantics and the terms are interchangeable, so I'll use *summary* from here on.

Pete 6/13/13 1:38 PM
Comment: Did you notice that I am parallel throughout the headline and ensuing sentences in that abstract is always mentioned before summary?

Write your summary only after you've written the main document, when you understand the subject most completely. Kick back, think, and ask yourself: What do my readers need to know if they have only a few minutes to understand what I've written? How and why does the document concern my readers? What's the bottom line for them and for me? What will readers take away—how will they be influenced--after reading the summary? Your answers are the focus and core of your summary.

The three types are 1) Informative (What the original contains; a recitation of content that is usually less useful to readers since the content is often outlined in the table of contents, and is not intended to be persuasive). 2) Descriptive (A discourse that extends beyond content and into implications and/or conclusions). 3) Editorial (Your reasoned opinions about the subject and its presentation).

Pete 6/13/13 1:38 PM
Comment: The three types of summaries are pretty much the same as the three types of heads, subheads, footnotes, and endnotes—a type of parallelism. Note that I created a horizontal list to demonstrate the format. I could have created a vertical list.

Note: I could have structured the previous paragraph as a list: The three types are:

1. Informative ...
2. Descriptive ...
3. Editorial ...

Note also that I used ellipses (...) to indicate that I omitted some information, and that I could have placed this entire structural option in a footnote.

Summaries are typically 150—350 words; however, length varies widely with the complexity and length of the original, and, at times, with the number of important audiences and their diverse interests. A longer, clear summary is always better than a shorter, murky summary. Think clarity and concision.

Summaries are located on the first page of the document, or right after the table of contents, to enable readers to decide if they will read the piece or not. They also can be placed at the end of a document to remind readers what they read and help them remember key points, often labeled *Conclusion.* If you write two summaries that essentially cover the same points, change your writing style.

To summarize summaries:

- Being aware of these imperatives will create maximum impact.

1. Find and stick to the essential message(s), the sum of the significant materials.
2. Write an independent, stand-alone, separate document structured as intro—body—conclusion similar to the structure of a good paragraph.
3. Adopt a style/tone suitable for receivers.
4. Be true to the original. If you editorialize, say so with italics, parentheses, or direct admissions.
5. Be concise; edit out unneeded words, phrases, sentences.

- Adhering to this process will enhance your productivity.

1. Read and understand the entire original.
2. Re-read and underline the main/essential points, then ignore all but the key phrases.
3. Connect and rewrite the key phrases; include all even if the piece becomes too long.
4. Edit for clarity and concision: convert passives to actives and nouns to verbs whenever possible, remove deadwood phrases that don't contribute to the message, and check for coherence and unity using the one-word outline.

- Staying within these guidelines will increase readership and understanding.

Pete 6/13/13 1:38 PM
Comment: Note the parallel use of the same verb form at A, B, and C, and at 1,2, 3 etc. And note that A, B, and C are editorial, while 1,2, 3 ... are instructional.

1. Put it all on one page to meet readers' expectations and for easy distribution.
2. Avoid technical jargon unless your audience is limited to those who understand it, which is rarely the case. Consider the limited technical proficiencies of many purchasing agents and upper managers, write more than one summary if necessary, and differentiate the summaries by using italics or bold typefaces for one and regular type for the other.
3. Include only important data, i.e., the data needed to meet your and readers' purposes.
4. Avoid references to the documents' body; they do not add to readers' understanding and are assumed to be there.
5. Create paragraphs just as you would in the main document, and use lists to shorten and highlight.

3. Know and stay within your structure.

The primary structures of documents are: general-to-specific (the most common that is also called deduction, whereby conclusions and results are presented first, supporting details next; this book is an example); specific-to-general (induction, selected often because it follows the normal chronology of a research or engineering project, but not necessarily the report); and chronological (to describe a process or occurrence; also called functional sequence).

Other structures include: effect-cause, in which an event is traced back to its apparent origin or cause, often confused with correlation. Example: When a ramp in the convention center collapsed, the cause was traced to a weak bolt. The opposite structure, cause-effect, merely reverses the order: A weak bolt caused the ramp to collapse.

Pete 6/13/13 1:38 PM
Comment: Note that I capitalized the first word after the colon when it starts a complete sentence, but not when it doesn't.

Pete 6/13/13 1:38 PM
Comment: Amazing how active voice cuts words and adds clarity.

4. Use transitions/connecting words and phrases

Inexperienced writers tend to shun transition words and phrases, thinking that they are useless, distract receivers from the message, and sound flowery. Not so: Transitions pave the way to the next thought and alert receivers to the content of the subsequent sentence/thought, helping them to understand the senders' thinking and adding clarity and, if desired, a conversational tone.

To indicate an additional thought: again, additionally, along with, also, too, and, as well as, besides, furthermore, moreover, next, together with, what's more.

Cause/effect: accordingly, and so, as a result, because of, consequently, due to, hence, thus, therefore, since.

Comparison/contrast: but, conversely, equally, however, in contrast, in the same way, similarly, on the whole, on the other hand.

Conclusion: all in all, altogether, finally, in brief, in short, to conclude, to summarize.

Condition: although, depending, even though, granted that, if, provided that, to be sure, unless.

Emphasis: above all, after all, again, as a matter of fact, indeed, to repeat, surely, unquestionably.

Illustration: for example, for instance, in other words, in particular, to demonstrate.

A paragraph in which the writer has used transitions properly:

Advertising a product on the radio has many advantages over television. **For one,** radio rates are much lower. **Specifically**, a one-time, 60-second spot on local TV can cost $700, about the same as nine 30-second spots on radio. Just as attractive are the low production rates for radio. **Still another advantage** for radio is immediate scheduling. **Often,** the ad is aired during the same week the contract is signed. **On the other hand,** TV programs are frequently booked months in advance. **Furthermore,** radio reaches more potential buyers. **After all,** radio follows listeners everywhere—in their homes, offices, and cars. Although TV is very popular, it can't do that. **All in all,** radio is an efficient medium for many advertisers. (The writer could have started the last sentence with: When all is said and done ...On the whole ...To summarize ...)

5. Replace text with visuals when feasible

We all know that, thanks to TV and other influences such as advertising, we live in a world in which visuals have reached new importance in all our communications. "A picture is worth a thousand words" is more valid than ever. By visuals, I mean graphs, charts, drawings, photos and so on—all can replace text and play nicely to your need for concision.

Guidelines for visuals are similar to the guidelines for prose: content—structure—and tone rear their prominence once again.

Content: be certain that your visual:

1. Serves legitimate purposes (clarification, not mere ornamentation, and concision).
2. Is appropriate for your message and for receivers' levels of knowledge and understanding.
3. Is easy to interpret: it is titled and numbered, accurate, explained in captions if needed.
4. Is introduced, discussed, and integrated into the text so that text and visual complement each other to create complete thoughts.

Pete 6/13/13 1:38 PM
Comment: I selected three types of bullets to demonstrate simple design choices, knowing that I am out of parallel. I would never do that in a document. Is my list a visual of sorts? I

Structure and design: be sure that your visual is:

- ✓ Easy to locate on the page and in the document.
- ✓ Positioned for balance on the page, and set off by white space.
- ✓ Spacious and uncluttered.

Tone: select the:

- Best type of visual for your purpose.
- Patterns, colors, and shapes that engage receivers.

Pete 6/13/13 1:38 PM
Comment: Did I need 'type of'? Delete it and decide.

Pete 6/13/13 1:38 PM
Comment: I veered out of parallel to demonstrate how easy it is; I can fix by: Visuals that meet your purpose, and Patterns, colors, and shapes that engage receivers.

6. Avoid dangling/misplaced modifiers and unclear pronoun references

A modifier is a word, phrase, or dependent clause (not a complete thought) that qualifies or adds meaning to other parts of the sentence; it must be clearly connected to what it modifies. When it isn't, it can change the meaning and destroy clarity. Here are some simple and humorous examples (replete with questions you might ask):

1. *After climbing the mountain, the view was beautiful. (Who climbed the mountain?)*
2. *Moving to Arizona, his rheumatism quickly improved. (Who moved? Can rheumatism improve?)*
3. *Dr. Jones instructed to patient while in the hospital to watch his diet very carefully. (Whose diet?)*
4. *Having been well-coached by his campaign manager, the candidate's speech added support for the candidate. (Who, what was well-coached? Note the hyphen. Needed?)*
5. *Blinded by the approaching car's headlights, John's car swerved off the road into a ditch. (Who was blinded?)*
6. *After opening up the closet, a cockroach ran for the corner. (Smart cockroach?)*
7. *The policeman shot the thief with a knife. (Impossible?)*
8. *Shoes are required to eat in the cafeteria. (Whose shoes can do that?))*
9. *Mia, the star of Team America, soothed the ankle she sprained in a scrimmage with an ice pack. (How did she sprain her ankle? Can you scrimmage with an ice pack?)*
10. *What can you do when your child is sick and you have to work? Call ABC for sick child care in your home. (What kind of care?)*

You can correct dangling/misplaced modifiers and unclear references by rewriting the sentence and/or rephrasing the modifier. I'll rewrite the ten examples and add a few comments; I hope that you approve.

- After I climbed the mountain, I was awed by the beautiful view.
- When John moved to Arizona, his rheumatism pains subsided quickly.
- While doing his rounds in the hospital, Dr. Jones advised his patient to stick strictly to the hospital diet. OR: Dr. Jones, while on his hospital rounds, advised his patient to not eat anything other than food served by the hospital.
- The candidate's speech supported his run for the office, largely because he was well-coached by his campaign manager. OR: The candidate—well-coached by her campaign manager—gave a rousing speech that is sure to help her win the election.
- Joe, blinded by an approaching car's headlights, lost control of his car and drove it into a ditch. OR: Joe swerved his car into a ditch, claiming that he was blinded by an approaching car's

Pete 6/13/13 1:38 PM
Comment: A compound sentence with two independent clauses, each with the same subject: I.

Pete 6/13/13 1:38 PM
Comment: I changed the tone dramatically.

headlights. OR: Joe claimed that, because he was blinded by an approaching car's headlights, he swerved into a ditch.

- When Pete opened the closet, he saw a cockroach run out and into a corner of the room. OR: When he opened the closet, Pete was startled by a cockroach as it ran into the room and into a corner.
- The policeman shot the thief, who brandished a knife. OR: The thief, who was wielding a large knife, was shot by the policeman in self-defense.
- Only people who are wearing shoes will be served in the cafeteria. OR: People who are not wearing shoes will be asked to leave the cafeteria.
- Mia, the star of Team America, soothed, with an ice pack, the ankle she sprained in a scrimmage. OR: Mia, the star of Team America, sprained her ankle in a scrimmage; she soothed it with an ice pack.
- When your child is sick and you have to work, we at ABC will care for your child in your home. OR: Call ABC for expert care of your sick child while you are at work.

Pete 6/13/13 1:38 PM
Comment: A good spot for passive voice?

Pete 7/23/13 9:19 AM
Comment: Is the reference to ankle clear?

CHAPTER V: INTERNALIZE SEVEN TECHNIQUES FOR CHECKING QUALITY

1. **Read aloud.** If you stumble over an awkward sentence or word, change it; if you run out of breath before you reach the end of a sentence, shorten it; if you question the clarity or meaning of a sentence, rewrite it. Trust your instincts.

2. **Record the document and play it back;** you'll be amazed at the blips you'll uncover by listening to yourself.

3. **Ask a trusted--and *unbiased*-- colleague to read your words.** A professional sender is your best sounding board, your best friend or spouse your worst. Be sure they know your receiver and purpose before asking for critique.

4. **Use the tools on your computer, but don't trust them**. Spellcheck is wonderful, but it can't identify a wrong word—one that conveys a wrong meaning--that is spelled correctly; only your mind can do that. Ditto grammar check. The app that highlights passive voice enables you to change to active when appropriate; use it.

5. **Distance yourself physically from your words**. Pin the document on a wall and step back, for example; you'll be amazed how you can examine your words more objectively.

6. **Fantasize, again, how your receiver(s) will react.** You can't do this too often.

7. **Use the following self-editing checklist:** Self editing has often been dubbed the sender's toughest job; this checklist may help: (in an alpha list, not numbered. OK?)

DO:

a) Be sure the receiver is given all necessary information. Complete the Receivers' Profile in Appendix I and adhere to it.

b) Guard against factual errors and a tone that is offensive in any way.

c) Correct all errors in grammar and syntax.

d) Conform to your organization's style manual or other published authority.

e) Correct all ambiguities and muddy meanings such as those caused by misplaced or dangling modifiers.

f) Edit for clarity, concision, and purpose by applying all the techniques you've learned.

g) Stick to the structure and format.

DON'T:

I. Make changes for the sake of change (elegant variation).

II. Overuse jargon. As useful as jargon can be for concision, it may not be familiar to all readers.

CHAPTER VI: INFUSE THE TRAITS OF SUCCESSFUL LEADERS

Communication and leadership skills are linked so tightly as to be one: Managers cannot lead if they cannot clearly articulate their vision. And communications skills are linked to promotion: Technical and quantitative skills soon give way to communications skills in people whose careers take off.

Historians, consultants, business managers, and many others have tried to define in just a few words the traits of leaders, whether they're in government or business. Paul Johnson, a historian and keen observer of the human condition who has written several best-selling books, condensed leadership to five traits: *Ideas and beliefs,* the central principles which the leader espouses passionately; *Willpower*, an unshakable confidence of correctness; *Pertinacity,* the patience and primitive doggedness to stick to the strategy; ***Ability to communicate***, to explain objectives and strategy in ways that implementers understand and believe in; and *magnanimity,* the greatness of soul that is so difficult to define but which is so apparent in those who own it.

Pete 6/13/13 1:38 PM
Comment: Would you prefer five vertical bullets?

Successful leaders have also been described as magnanimous, humble, prudent, courageous, principled, fair, and visionary. They can exhibit these traits by ***aligning their actions and words that are expressed clearly, concisely, purposefully, and truthfully***.

In his book, *Leading Minds: An Anatomy of Leadership,* Howard Gardner describes a continuum of leadership: Indirect, exerted through scholarly work or symbolic communication, and direct, exercised through speeches and other communication. I interpret that to mean that leaders understand the power of their words to influence others, and exploit that power to its fullest.

CHAPTER VII: APPLY WHAT YOU'VE LEARNED BY EXAMINING THESE BEFORE-AND-AFTER EXAMPLES

1. A few simple snippets to sharpen your s**t detector:

Taking a vitamin D supplement might reduce the chance of dying for women older than 70, but only if it's vitamin D3 ...

Eat breakfast within 90 minutes of waking, every morning. A meal that has protein, complex carbohydrates, and healthy fat has been linked to better cognitive performance and a brighter mood during the day.

I am an elderly woman who needs roof repairs.

ABC—A powerful resource for industry and governments.

Pete 6/13/13 1:38 PM
Comment: But everyone's chance of dying is 100%.

Pete 6/13/13 1:38 PM
Comment: Where's the attitude?

Pete 6/13/13 1:38 PM
Comment: The first and second thoughts are unrelated, a sin of thinking called non sequitur.

Pete 6/13/13 1:38 PM
Comment: Do you suggest a wig?

Pete 6/13/13 1:38 PM
Comment: Out of parallel: *industry* is a collective noun, *governments* a plural noun. Try *industries and governments.*

2. An incoherent paragraph written by an energy engineer:

Solar power offers an efficient, economical, and safe solution to the Northeast's energy problems. Unlike nuclear power, solar power produces no toxic waste and poses no danger of meltdown. Solar power is efficient. Solar collectors could be installed on fewer than 30 percent of roofs in the Northeast. These collectors would provide more than 70 percent of the area's heating and air conditioning needs. Solar power is safe. It can be transformed into electricity. This transformation is made possible by photovoltaic cells (a type of storage battery). Solar heat collectors are economical. The photovoltaic process produces no air pollution.

Pete 6/13/13 1:38 PM
Comment: Note that I've continued the format to italicize *before* examples and then revert to regular type for after. I'll stay with that format for the remainder of the book—to be parallel.

Note: The topic sentence expresses an attitude and sets readers up to read about efficient, economical, and safe in that order. Then the body wanders to safe, efficient, efficient, economical, safe, function, function, economical, safe. The sender cooked a stew of incoherence/disunity that chunking and bulletizing can reverse:

Solar power can help solve the Northeast's energy problems. Solar power is:

1. **Efficient:** collectors installed on 30 percent of the area's roofs could provide 70 percent of the energy needed for the area's heating and air conditioning.
2. **Economical:** collectors operate for up to 20 years with little or no maintenance, accruing savings that enable initial costs to be returned within ten years.
3. **Safe:** collectors don't emit any air pollution (unlike combustion of coal, oil, or wood), and they don't produce toxic wastes (unlike nuclear power).

(88 words, down 13 percent from the original.)

Another topic sentence?

Solar power can help meet the Northeast's energy needs because it is:

(The bullets above)

Can we change the last bullet?

- **Safe and environmentally benign:** collectors don't pollute the air (as combustion of carbon-based fuels does), and they don't produce toxic wastes (as nuclear plants do).

3. A major consulting firm describes its business:

Since 1989, companies have turned to ABC as a trusted partner. We continue to work on the most challenging and interesting projects delivered by exceptional people.

Founded in 1989, Actual Big Corp (ABC) is a company of professionals that embrace three core principles: Senior Leadership, Integrated Services, and Personal Business Relationships. ABC's main focus is on the partnership that we develop and maintain with our clients and business partners. Operational flexibility and the needs of business are key factors in ABC's commitment to fiscally responsible project management.

ABC has a proven track record of providing sound, technological solutions that balance the requirements of compliance and risk. ABC works in the primary practice areas of environmental, civil, and site development engineering, ecological, waste management and water resources. These technical consulting and engineering services are delivered to private industry, real estate developers, architects, waste management, legal and public sector clients

Let's bulletize and add thoughts to complete the abstractions:

ABC (Actual Big Corp) embraces three core principles:

- Senior leadership: Professionals with an average of fifteen years experience in their respective fields lead project teams.
- Integrated services: ABC's five practice areas—civil and site development engineering, ecological services, environmental services, waste management, and water resources management—complement each other on many projects.
- Personal business relationships: The satisfaction of our clients, and incidentally ABC's growth, is a direct result of the close and lasting

Pete 6/13/13 1:38 PM
Comment: The first and second sentences are unrelated, another non sequitur.

6/14/13 6:22 AM
Comment: Noun/verb agreement

Pete 6/14/13 6:23 AM
Comment: A perfect example of incoherence, aka unconnected, jumbled thoughts.

6/14/13 6:26 AM
Comment: List is way out of parallel.

6/14/13 6:26 AM
Comment: Do the second and third sentences support the topic sentence?

6/14/13 6:27 AM
Comment: I changed the order of information in parenthesis; either order is correct.

relationships developed among client and ABC personnel at all levels.

The three core principles are complemented by a strong focus on clients' needs via operational flexibility and responsiveness to business objectives and regulatory requirements.

6/14/13 6:29 AM
Comment: Notice how I connected thoughts so the receiver needn't, and I put the receivers' interest first.

ABC serves a broad range of clients in the public and private sectors: Governments at all levels, authorities, architects, real estate developers, attorneys, and others.

4. A doctoral candidate wrote his thesis for two sets of receivers: top, very knowledgeable management of a large engineering company interested in starting a new business, and the smart but less-business-savvy review committee at a university. He opened his thesis with this:

Along with the economic changes, the large utility boilers that are utilized in the power generating industry are highly regulated from an environmental perspective. The electrical power producing industry is the largest contributor of pollutant emissions to the air and water. Because of this, this industry is highly regulated at the state and federal levels.

6/14/13 6:30 AM
Comment: Start of a string.

6/14/13 6:33 AM
Comment: The first sentence is not related to the second, creating a *non sequitur* and disunity. The third sentence repeats the first.

You can rewrite in several ways:

a) Electricity generators emit more air and water pollutants than any other industry. Therefore, they are regulated heavily at State and Federal levels.

b) Electricity generators are the nation's worst polluters of our air and water. It's not surprising, therefore, that they are regulated heavily by State and Federal governments.

c) The emissions from power plants are regulated heavily by State and Federal governments. The reason: The plants emit more pollutants to our air and water than any other group of industrial facilities.

d) Power plants defile (foul?) our air and water more than any other group of industrial facilities; therefore, and not surprisingly, such plants are regulated heavily and monitored stringently by State and Federal governments. [Or, Therefore, State and Federal governments are justifiably regulating them heavily and monitoring them zealously.]

6/14/13 6:33 AM
Comment: I changed the tone from benign to aggressive.

5. A letter from a building manager to resident s to alert them to extraordinary high costs of water:

Recent invoices from XXX for your building was exceptionally high compared to past bills.

As you know, everyone shares in the actual cost for water and sewer usage. Each homeowner has a responsibility to ensure that they are not wasting water. Please check all your water supplies and repair all leaky faucets and toilets immediately.

Thank you very much for your cooperation.

6/14/13 6:34 AM
Comment: Invoices was? C'mon..

6/14/13 6:35 AM
Comment: If I know, why are you telling me?

6/14/13 6:37 AM
Comment: Responsibility is a noun that could be an adjective, responsible, deleting a few words and adding a touch of action.

6/14/13 6:38 AM
Comment: The subject, *Each*, is singular; they, referring to *Each*, is plural.

Let's revise to better meet the purpose, but in passive voice:

Recent invoices from Pittsburgh Water Authority for your building were higher than in the past and for other buildings. Therefore, please repair any leaking faucets and toilets, the most likely causes.

Your efforts will lower costs for all residents. Thanks for your cooperation.

(43 words)

Another way that meets the purpose more quickly; did I write in active voice?

Please check all your faucets and toilets for leaks and repair them; they are the likely causes of extraordinarily high charges from the Pittsburgh Water Authority for your building. Your efforts will not only lower your costs, they will also lower costs for all residents.

6/14/13 6:39 AM
Comment: The pronoun *they* refers to *efforts*; is that clear?

(45 words)

Still another way:

I'm concerned that the invoices for water in your building are extraordinarily high, raising costs for all residents. Therefore, please check etc

6. A letter from a real estate agent in Florida. A bit of background: I was planning to buy a condo in Florida for two atypical purposes: to profit financially (I am a knowledgeable and

ambitious investor) and to get away from my business and indulge my tastes for jazz and good food (I hate sun, sand. and golf). I called a real estate agent, explained my purposes, and received the following letter:

Dear Mr. Geissler,

Enclosed is the pertinent information regarding the condo unit we discussed. The complex is located on the Gulf of Mexico. This particular unit faces the waterway opposite the Gulf. This waterway is known as the Intercoastal Waterway. The view in this direction is no less beautiful. Many people prefer this view over the Gulf.

6/14/13 6:40 AM
Comment: The topic sentence is passive and lacks attitude.

6/14/13 6:41 AM
Comment: Than what?

As is explained in the brief, this penthouse was originally offered for $239,000. It has since been reduced at least twice. According to market comparables and sales in this complex, it is our opinion that the current listing value is between $240,000 and $250,000.

If you are interested, please contact us. We look forward to hearing from you.

In 123 words, this sender wasted his and my time—dropping productivity and raising costs-- by not telling me what the unit costs, why it's a good investment, or where the jazz clubs are. Guess what? The sender lost an easy commission and I lost a potentially lucrative investment and good times. I deposited the letter and its attachment in the nearest dustbin while muttering "idiot". Here's what he might have written:

Dear Mr. Geissler,

Real estate in this area has been a lucrative investment for decades, and we see signs that it will continue to be. . Since 1980, for example, selling prices have appreciated fourteen percent/year, three points faster than the DJIA. We are convinced that last year's growth of only one percent is an aberration caused by the nationwide recession, and appreciation will recover to historic rates very soon. Therefore, the slowdown opens a rare opportunity to buy: The asking price of the condo we recommend has been reduced from $200,000 to $170,000, and interest rates for mortgages are below five percent.

Short-term rentals can add to the investment value of your property. Vacationers are always looking for clean properties located in the area, and rents can be as high as two thousand dollars/week.

We suggest a condo for you that is located only a five-minute walk from the cultural district in Clearwater Beach, in which three popular bistros feature local and nationally recognized jazz groups. In addition, two five-star restaurants in the neighborhood are ready to serve you in quieter settings.

In summary, real estate in this area has been, and very likely will be, an outstanding investment. The jazz and other entertainment in the area are equally outstanding.

Please review the attached documents for details of the property, and call me with any queries. Thanks for your interest.

226 words, but they clearly and concisely met my purposes—and had a good chance of meeting the agent's purpose of selling a property and earning nice commission.

7. A marketing manager for a major manufacturer wrote this memo to help the CEO decide if a plant should be closed. The CEO bounced the memo back with a terse note asking for clear direction, painting my friend as stupid. He came to me and asked why the memo failed.

The plant in Philadelphia can produce three million, 600 thousand pounds/year, while the capacity of the plant in North Carolina is 2100 tons per year. In light of market demand, the smaller plant can be closed. (39 words)

Try this:

The plant in Philadelphia can produce 3.6 million pounds/year, 15 percent less than the plant in Charlotte, which can produce 4.2 million pounds/year. In light of market demand, which is projected at 4.0 million pounds/year, the plant in Philadelphia can be closed. (41 words)

Or, to write the last sentence in active voice: The plant in Philadelphia can produce 3.6 million pounds/year, 15 percent less that the plant in Charlotte, which can produce 4.2 million pounds/year. We project market demand to be 4.0 million pounds/year, and recommend closing the plant in Philadelphia. (39 words)

Or, to reverse the ideas: I recommend strongly that we close the plant in Philadelphia: It can produce only 3.6 million pounds/year, 400,000 pounds short of market demand. The plant in Charlotte, however, can comfortably meet demand by producing 4.2 million pounds/year.

8. The CEO of another large firm sent this memo to 2300 employees to ask them to not use certain parking spaces and cafeterias during a meeting attended by more than 300 guests. Empathize with employees/receivers and ask yourself if the memo met its purpose:

SUBJECT: MEETING ON DECEMBER 10—12

On December 10—12 the company will host a conference which will be held in Building Six auditorium.

Approximately 300 conference participants are expected, which will cause crowding at some of the company's facilities. The Building Six parking lot will be reserved for conference attendees. Employees who normally park there should make an effort to use other lots. Also, between 1130a.m. and 1230p.m. the Administrative Building cafeteria will be occupied by conference attendees. Company employees should plan accordingly.

As always, your cooperation in these matters is greatly appreciated.

Ugh! Eighty-eight words in passive voice; the point is buried at the end of the second paragraph; and every employee had to read the memo to find out if it meant anything to him or her. Productivity took a dive. Try this:

TO: Employees who park in the Building Six lot and/or eat in Cafeteria Three.

Please use other parking lots and cafeterias on December 10 through 12 so that some 300 guests can use them while they attend a conference here.

Thanks for your cooperation.

Too terse at 44 words? Maybe, but it made the point in a hurry.

9. This one page proposal for training personnel in corrosion protection is the

absolute poster-boy for bad writing. Please use it as a final exam; I will point out the plethora of sins with numbers in parentheses; you decide, before looking at my rewrite, what they are and correct them in your mind or, preferably, jot them in the margin, then rewrite.

Dear Customer,

In accordance with our electronic mail correspondence on January 3, 2010, (1) supplier is pleased to present customer with a proposal (2) (3) for coatings training services. (4) Supplier is the oldest, largest and most experienced coatings consulting/engineering company in the United States. We have been training coatings suppliers, manufacturers, applicators and end users since 1968, and we are confident that our training services can benefit (5) your company as well. Because this is a closed course conducted solely for customer personnel, the content can be tailored to cover the areas you fell (6) will be most beneficial. It (7) is proposed that the course content be essentially identical to that attended by Mr. Smith earlier this month, except that hands-on blast cleaning and coating application may need to be eliminated from the curriculum, unless equipment, facilities and personnel can be supplied by customer. (8)The proposed modules are listed below in the order of presentation:

- *Introduction to Coatings and Coatings Inspection*
- *Corrosion Theory and Prevention*
- *Corrosion Types and Characteristics (focus on the coating types used by customer)*
- *Surface Preparation Equipment and Techniques and Standards*
- *Coating Application Equipment and Techniques; Comprehending Product Data*
- *Coating Inspection Techniques/Instrumentation*
- *Coating Failures*
- *Course Examination (9, 10, 11 for this list)*

Workshops conducted throughout the course enable participants to apply the classroom theory, and to obtain immediate feedback as to whether the information conveyed was learned (12). Other topics can be added substituted as desired. (13) The list of proposed topics will consume approximately 2 and a half days. (14)

Cost (15)

The coatings training course will be provided at your facility in Wisconsin. (16)The cost of the course will be $6700 plus travel expenses *for two instructors and shipping charges, which will be invoiced at cost. There is a $15.00 per person fee for training materials.* (17)

Customer will responsible for providing a suitable training facility, a 35mm slide projector with wireless remote and a flipchart and markers for the course. (18) All training supplies, instrumentation and participant workbooks will be supplied by supplier. (19)

If you would like supplier to conduct the course as described herein, (20) please acknowledge acceptance of this proposal by signing and returning the attached authorization to proceed. The course can subsequently be scheduled. (21)

We look forward to working with customer on this training endeavor. If you have any questions regarding the information contained herein, call XXX

Here we have 385 words with at least 21 sins that create confusion and obfuscation. Try this:

Subject: Training of ABC personnel in corrosion and its prevention.

The following proposal is founded on our mutual interest in preventing corrosion, as expressed in our emails of January 3, 2010:

The content[4] of the 2.5-day course will be configured to meet your most pressing needs by examining the following topics in this sequence:

1. *Overview:* what coatings are and aren't; coating types; understanding product specs; techniques for inspection.
2. *Theory:* corrosion and its prevention
3. *Practice:* demonstrations and hands-on workshops of equipment/techniques/standards/installation for preparing surfaces, applying and inspecting coatings, and analyzing failures.
4. *Exam*

The course will be held at your facility in Madison, WI, in which you will provide a 35mm slide projector with wireless remote, a screen, and a flip chart with markers.

Our company's outstanding 34-year record of training and prevention has made us the largest and most experienced in the field, Clients

[4] Content will be similar to the course attended by your Mr. Smith without hands-on demonstrations of blast cleaning and coating applications, unless you supply suitable equipment and facilities.

report that corrosion at their facilities has diminished and costs have fallen.

Your instructors are fully certified: Joe Jones, for example, is certified by two industry groups, and Sam Smith by three. Please refer to the complete resumes attached.

Your costs will be $6700, and includes training materials for 20 attendees, travel for two instructors., and class time.

Your next step is to sign and return the attached authorization, after which we will work with your representatives to finalize the schedule and agenda.

Thanks for your consideration.

(Signed)

(268 words, down 32% from the original)

10. An engineering firm sends a request for a proposal to another engineering firm:

RE: Price Quote for HVAC Evaluation

Dear Mr. X,

As we discussed last week, ABC Company is bidding to perform work for a client located in Meadville, PA. This work includes evaluation of the HVAC system. Our client has experienced "apparent deficiencies in indoor air quality and control of ambient air temperature in two leased spaces." These deficiencies have resulted in an unsuitable work place environmental and may also have contributed to acute and chronic illness in certain employees over a seven year period." (From the client's letter of January 3, 1995.)

The two leased locations are both located in a single story retail outlet mall. The first location is an approximately 8000 square foot area that is divided into individual professional offices. The second location is separate from above and consists of a 1600 square foot area divided into offices and clinic rooms and is utilized by health professionals.

Scope-Of-Work

1. *Evaluate the existing HVAC system in the above locations, specifically:*

a) *Adequacy of airflow to each individual office, work room and common use area.*
b) *Adequacy of fresh air dilution to each individual office, work room and common use area.*
c) *Adequacy of ambient air temperature control in each individual office, work room and common use area.*
d) Indoor air quality to include the following indicators: carbon dioxide, carbon monoxide, ozone, radon and particulates.

2. *Review a recently proposed solution to correct certain obvious deficiencies in the existing system and indicate the adequacy of the proposed changes in:*
   a) *Providing compliance with conditions of the current lease.*
   b) *Providing adequate indoor air quality and temperature control.*

3. *Recommend system changes to resolve or abate any and all HVAC and indoor air quality problems disco ' ' ·ing the evaluation process.*

I know we had discussed meeting to review this project; however, the client has decided that he would like to receive all bids by January 23. I believe this study will be used to negotiate a new lease from the facility owner. The review required in Scope-Of-Work #2 is a proposal submitted by the owner to correct the tenants complaints.

Please get me a price ASAP. Your price should be all inclusive: technician, engineer, equipment, materials, travel, report and communication. We would prefer a lump sum cost rather than a price schedule. Call if you have any questions or it you decide not to submit a price. We look forward to a continuing relationship with your firm.

Sincerely,

ABC Consulting
Operations Manager

412 words

Dear Mr. __________,

As discussed last week, we will submit our proposal for evaluation and certain other work related to an HVAC system owned by a client in Meadville, PA, and would like you to be our partner. The work is necessary since the client has experienced “apparent deficiencies in indoor air quality and control of ambient air temperature in two leased spaces,” rendering the workplace unsuitable and possibly contributing "to acute and chronic illnesses in certain employees"* during the past seven years.

Description of Sites

Both leased sites are in a single-story retail mall: One is 8000 square feet of offices; the second is 1600 square feet of offices and physicians' examination rooms.

Scope of Work

1. Evaluate the capabilities of the HVAC system to provide, for each office, work room, and common use area, adequate:
   - airflow,
   - fresh air dilution, and
   - ambient air temperature control.

2. Evaluate air quality, as measured by levels of carbon dioxide, carbon monoxide, ozone, radon, and particulates.

3. Review and evaluate the adequacy of solutions proposed recently by the owner to correct certain deficiencies, including their potential to comply with conditions of the current lease and create a comfortable, healthy environment.

4. Recommend changes to the HVAC system that will resolve or abate any problems uncovered during evaluation.

Price/Schedule

Please submit your lump sum price, to include all costs of manpower, materials, equipment, travel, reports, and communications, before January 23, just before tenants will negotiate new lease with the owner. (Footnote?)

Please call with any questions, or if you decide to not submit your proposal.

We look forward to working with you again.

Sincerely

---

* From the client's letter of January 3, 2010.

(284 words, down 31%)

11. From a doctoral thesis:

The electric utility industry is divided into three segments called generation, transmission, and distribution. Historically, typical electricity consumers have had little reason to be familiar with these terms or the structure of this industry. With the advent retail competition, however, consumers will need to understand these terms to reevaluate the new choices becoming available to them.

Generation is the process by which fuels (gas, coal, nuclear fuel, etc.) or renewable sources of energy (solar energy) are converted into electric energy. Transmission is the process by which the generated electricity is moved in bulk from the generation plant to the wholesale purchaser. Distribution is the process of delivering the power from the wholesale purchaser to the retail consumer.

The reason electricity consumers need to know these terms are that they are now able to buy each of these services from more than one company. The program in place for Pennsylvania residents allows the consumer to choose the generator of their electrical power. The transportation and distribution segments will remain with your current provider and the consumer cannot choose who provide these services. Examining this in light of the previous discussed environmental regulations, the electricity generation industry is experiencing pressure from two fronts. This creates an opportunity for ABC Inc. to evolve and offer their current services as a package solution rather than as individual services.

(224 words)

The electric utility industry is typically divided into three functions:

- Generation, to convert into electricity fuels such as coal, oil, natural gas or uranium, or renewable sources of energy such as sunlight.
- Transmission, to convey power in bulk to the wholesale purchaser.
- Distribution, to deliver power to the consumer.

Consumers have had little reason to understand these functions until deregulation spawned competition, and, in turn, the opportunity to purchase power from any number of suppliers. In Pennsylvania, for example, consumers can purchase power from the generator of their choice, but not the transmitter or distributor.

Competition, combined with the environmental regulations discussed above, pressures the industry on two fronts (which are?), and opens opportunities for ABC Inc. to evolve and offer its individual services as packaged solutions.

(132 words)

Or, to replace the bullets with visuals: (should bullets be numbers?)

The electric utility industry is typically divided into three functions:

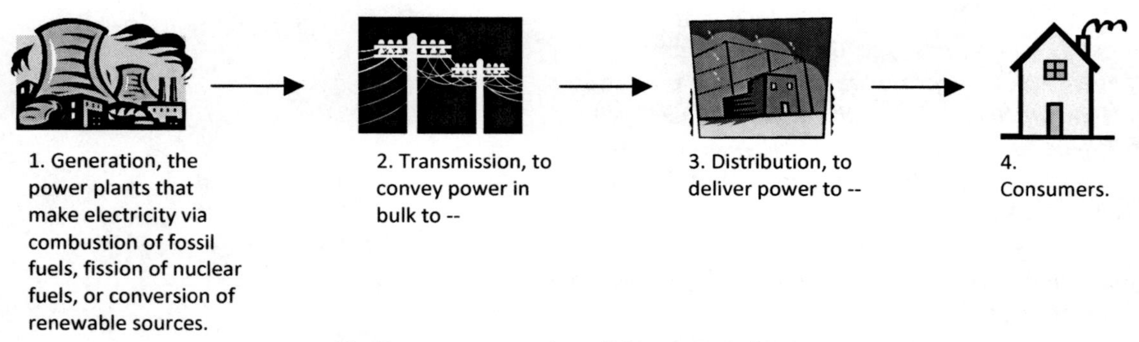

12. From a report describing the working conditions in a factory:

The purpose of this survey was to evaluate occupational exposures to airborne particulates and welding fumes associated with representative routine Metal Inert Gas (MIG) and Shielded Metal Arc (electric weld) welding operations and related pneumatic chipping or grinding and/or limited torch cutting of carbon and stainless steel grades at the Structural Shop, Field Forces Group. In addition to particulates, exposures to gases which may be emitted during the welding process were evaluated. The gases sampled were ozone, carbon dioxide, carbon monoxide, and nitrous fumes.

(84 words)

Purpose: Evaluate occupational exposures to airborne particulates and welding fumes (ozone, carbon dioxide and monoxide, and nitrous oxides?) associated with representative routine:

- Metal Inert Gas (MIG) and shielded metal arc (electric) welding; and
- Related chipping and grinding and/or limited torch cutting carbon and stainless steel at the structural shop, Field Forces Group.

(53 Words, down 37 percent)

CHAPTER VIII: REVIEW ALL THE TOOLS YOU NEED TO WRITE LIKE A PRO

ALWAYS

- Know your incentives, your immediate and long-term motives; good writing is the interim goal on the way to the three Ps: Promotability, Productivity, and Profitability. Good writing is the route to influencing receivers in ways that you, the sender, want.

- Know the four traits of writing that define "good" and write to achieve them: Clear (the intended meaning is understood; it cannot be misunderstood); Concise (the right number of thoughts/words needed to meet receivers' and your purposes); Logical (the sequence of thoughts makes sense); and On-Point/Purposeful (to focus the message and avoid extraneous excursions to irrelevancies).

- Practice the habits and techniques until they are habituated. They will raise your productivity and your sendings' quality: its clarity, concision, and purposefulness.

- Consult the dictionaries/handbooks on or near your desk, or in your computer, for any queries on punctuation or usage.

- Understand the writing/communications processes, including the four-step process of Think before you write, Think while you write and write backwards from the details to the abstract; Think after you write, and Think after receivers react. Your writing always reflects Your thinking/intelligence.; don't let it reflect your lack of either.

- Be your own toughest editor: you have the checklist.

- Punctuate properly and your writing will be clearer. Refer to the rules in this book and in your handbooks.

- Develop paragraphs and your abstracts, summaries, and entire documents in the ABC format.

- Write topic sentences with attitudes to sharpen your focus on the topic at hand.

- Use appropriate transitions to help readers understand your thinking.

- Read, think, and listen analytically; remember Hemingway's prerequisite for being a good writer.

- Create smooth, connected sentences by obeying the this-that-these-those-same thought rules, helping avoid choppies and run-ons.

ALWAYS BE

- Influential; it's your overarching purpose as defined by the Receivers' Profile.
- Parallel with your sentences, paragraphs, and complete documents; it brings order out of chaos as much or more than any other technique.
- Coherent, to display logical connections of thoughts and to avoid disunities.
- To the point; your receivers are busy and become cranky waiting or searching for the nut of your message.
- Within your structure, also to create a sense of order.

ALWAYS AVOID

- Muddy modifiers/references; watch those pesky pronouns.
- Run-on and choppy sentences to avoid confusing and/or boring your receivers.
- Strings and the meanings they confuse.
- Smothered verbs, to speed the dialog.

ALWAYS PREFER

1. Active voice; it reduces your word count by 20--30% and adds action and clarity.

USE WHEN APPROPRIATE

2. Passive voice; sometimes hiding the actor is good.
3. Lists; bullets and numbers have their separate uses, and they can be reversed to prose.
4. Visuals, to replace words and add clarity.
5. Footnotes and endnotes to sustain the main dialog.
6. Summaries/abstracts, especially in longer documents.
7. Heads, subheads to alert readers to the main ideas and to transition easily to another topic.

APPENDIX I: THE RECEIVERS' PROFILE

1. Primary receiver(s).
   Name(s) and title(s).

2. Secondary receiver(s), if any:
   Name(s) and title(s).

3. The purpose(s) of the document, i.e., the change in thinking and/or behavior that you
Want (to affect content, structure, and tone):

1. Key points (to affect content and structure):

   Main

   Supporting

2. Receivers know the subject and technology (to affect content, structure, and tone):
Well……………………….a little………………………….not at all.

3. Based on receivers' knowledge of the subject and technology, the tone should be:
Detailed……………………………………………………….general
Technical……………………………………………………..lay

4. Receivers' feelings toward the subject (to affect tone):
Hostile……….resentful…………………wary…………………accepting

5. Receivers' feelings toward the writer (to affect tone):
Hostile……….resentful…………………..wary…………………accepting

6. The judge's feelings toward the subject, your firm, and you
Hostile………resentful………………..wary…………………accepting

Pete 6/14/13 6:53 AM
Comment: Complete the profile and then take the time to contemplate how your answers shape content, structure, and tone.

Pete 6/14/13 6:55 AM
Comment: Do not fall into the trap of profiling your receivers as "general" or everybody; all sendings are targeted at a specific receiver or group.

Pete 6/14/13 7:36 AM
Comment: All the outline you need to start writing.

Pete 6/14/13 7:37 AM
Comment: You can use jargon and technical terms if your receivers know the subject.

Pete 6/14/13 7:38 AM
Comment: You aren't loved by everybody; be more conciliatory to those who aren't friendly.

Pete 7/23/13 10:59 AM
Comment: Avoid absolute words like *guarantee*?

APPENDIX II: TIPS ON GETTING TO THE POINT AND ENHANCING UNDERSTANDING

1. Identify your and the receivers' purpose(s) by asking yourself: What do you want receivers to do; what are the main points of interest/need, and are they shared by receivers and senders.
2. Study the structure (organization) and relationships of ideas: Does the intro prepare readers for what follows; are relationships among ideas drawn clearly so that readers do not draw their own; are transitions from one topic to another clear and logical.
3. Analyze content by asking: Do details—including examples and anecdotes--sufficiently support the main ideas or concepts?
4. Sharpen the style/tone: shorten sentences that are too long [run-ons] and connect those that are too short [choppies]; vary sentence length [the average is 20 words] and structure; be sure your words are appropriate to the purpose and subject; watch for connotations that may insult or embarrass receivers.

Pete 6/14/13 7:44 AM
Comment: Denotation is the dictionary definition; connotation is the emotional impact of a word.

APPENDIX III: THE TECHNIQUES OF INFLUENCING

The purpose of every communication in business and government –and I'd wager in personal interactions--is to influence: to create a desired effect, impression, or impact on receivers of messages, whether they are readers or listeners. The desired effect may be to buy your product, think highly of your skills and organization, promote you to a better job, and bestow an award, and so on. To be most influential, consider:

Your image/credibility: Receivers will be more likely to be influenced by senders they think of as like themselves than they think of as different. Engineers are hired to sell to engineers, writers to teach writing, and so on. Could Oprah sell a breaker (or a wall switch, for that matter)? A large design/build project? Surely not, but she can sell books and household items.

Your message's opening: The influence of your message is increased dramatically if it opens with views shared by receivers. Thus, a proposal for a new breaker might better begin with statements about its safety and financial benefits than its supporting technology.

Your message's closing: In general, explicit conclusions are more likely to influence than indefinite conclusions that enable receivers to connect the dots on their own. Exceptions are receivers, typically top managers, who resent others who attempt to make decisions they are paid to make. The tone of the message often overcomes that roadblock by couching explicit conclusions and recommendations as suggestions or considerations.

In essence, the influence of any message relies heavily on that old standby: Know your receiver(s)—his or her needs, wants, values, and interests—and address them with the content, structure, and tone of your message. Complete the profile.

And always remember…the most influential communications are those that are clear, concise, and purposeful…that adhere to the disciplines of good writing. Well-crafted, reasoned communications impress receivers with the sender's intelligence, while poorly crafted communications impress receivers with the sender's ignorance. The result is simply that, assuming other factors to be equal or similar, intelligence, expressed by good communications, wins the contract, job, the next RFP, and so on.

APPENDIX IV: THE SEVEN MYTHS OF WRITING

1. Write the way you talk. Your conversations are replete with grammatical errors and irrelevant wanderings off point. If you don't believe that, tape and transcribe a meeting and apply your newfound know-how about writing to the words on the page.
2. Start with an outline or executive summary. Instead, complete the Receivers' Profile and use it as a guide to be changed as you write. I found years ago that detailed outlines and summaries are a waste of time, particularly if they lock in a document's structure.
3. Nobody can learn to write; it's inborn. Let's call this a half-myth: good writers tend to have the ear for rhythm and sense of organization that bad writers lack, but that doesn't preclude anyone from developing and applying the habits and techniques that separate good from bad. Writing is as much attitude and understanding as it is skill.
4. Writing is easy. Yes, bad writing is easy, but good writing is very difficult. Consider that I can write 2000 words an hour that I'll never admit are mine; I can write only 125 words an hour that I'll send to my clients or publishers.
5. You write to communicate. Yes, up to a point. At least as important: You write to grapple with ideas to give them shape and bring them to your conscious mind, i.e., we write to help us think. Ergo, writing reflects—and is a permanent record of--your thinking.
6. Studying grammar will improve your writing. Yes, you need to know and follow the few rules of grammar that are in this book, but studying it deflects you from thinking about content, structure, and tone—worsening your writing..
7. You can multi-task. Nobody can do two cerebral tasks at once—our brains won't allow it. The operative word is *cerebral.* Yes, you can chew gum and walk at the same time, but that isn't multi-tasking for senders or receivers.

APPENDIX V: USEFUL ABBREVIATIONS

- **AKA, A.K.A., aka, a.k.a.,** all mean 'also known as', and can be written in several ways, as I just demonstrated. I used it in several ways in this book; you shouldn't in your documents. Instead, select a style and stick to it throughout—another form or parallelism. Aka seems to be favored today.
- **et al** means "and others" and is typically placed at the end of a list, usually of people, to indicate all those in the category: Pete spoke to Tom, Ann, Jesse et al. Do not place a period after et al unless it falls at the end of a sentence, as I did above.
- **etc.** means "and other things of the same kind", not people. Never write "and etc"; it's redundant. Place a period after etc. no matter where it falls in the sentence or phrase.
- **e.g.** means "for example". Never write etc or et al after e.g., and typically place a comma before and after e.g.: Pete wrote, e.g. , a million words on energy
- **i.e.** means "that is" and is followed by a list, phrase, or clause to clarify what preceded. Place a comma before but not after i.e.
- **Sic**: not my error; that's how it appears in the original. Sic points out a grammatical error, misspelling, misstatement of fact and the like. For example: *The dog chased it's (sic) tale (sic).*

APPENDIX VI: FURTHER READING

Bryson, Bill. 2002. *Bryson's Dictionary of Troublesome Words: A writer's guide to getting it right*. New York: Broadway Books. (As expected, Bryson converts a dry topic to laugh-out- loud humor; read all his books to discover and enjoy his cheeky style.)

Horton, Susan. 1982. *Thinking through Writing*. Baltimore: Johns Hopkins Press. (I love this book so much that I've offered to create a class in analytical thinking based on it.)

King, Stephen. 2000. *On Writing: A Memoir of the Craft.* New York, Scribner. (I consider this the absolute best and most entertaining book on writing that I know of. No preaching, no tedium.))

Kolin, Philip. 2004. *Successful Writing at Work*. New York: Houghton Mifflin. (This thick tome is an exhaustive treatment for those looking for the complete experience.)

Leary, William and Smith, James. 1951. *Think Before You Write*. New York: Harcourt Brace and Company. (An oldie but goodie that dissects writing as thinking as no other book does. I recommend it highly.)

Lovenger, Paul. 2000. *The Penguin Dictionary of American English Usage and Style.* New York: Penguin Reference. (Very helpful examples of usage, many by well-known writers.)

Marshall, Lisa and Freedman, Lucy. 1995. *Smart Work: The syntax guide for mutual understanding in the workplace.* Dubuque: Kendall/Hunt. (I love this headline in the book: Communication *is* the new work!)

Merriam Webster. 2000. *Concise Handbook for Writers.* (An indispensible tool for guiding and checking punctuation.)

Michalko, Michael. 1991. *Thinkertoys: A handbook for creativity for the 90s.* Berkeley: Ten Speed Press.
(Michael moves from concept to exercises that will open your mind to new ideas.)

Truss, Lynne. 2003. *Eats, Shoots and Leaves: the zero tolerance approach to punctuation*. New York: Gotham Press. (A best seller for a reason: It's huge fun.)

Pfeiffer, William. *Technical writing: A practical approach.* 1994. New York: MacMillan. (This tome is an OK mixture of concepts and demonstrations that cover the vast universe of writing.)

Weaver, Richard. 1948. *Ideas Have Consequences.* Chicago: University of Chicago Press. (Another oldie but goodie that I guarantee will create many aha! moments.)

Zinsser, William. 1989. *Writing to Learn: How to write—and think—clearly about any subject at all.* New York: Harper and Row. (Zinsser demonstrates with anecdotes how writing cleans and clears thinking.)

Zinsser, William. 1980. *On Writing Well.* New York: Harper and Row. (A perennial best seller, Zinsser has written a wonderful example of metadiscourse; no preaching, no tiresome rules and jargon.)

APPENDIX VII: BONUS: ADDITIONAL EXAMPLES FOR ADDITIONAL INSIGHTS

From a brochure

- *ABC Engineers, Inc. is a full-service firm specializing in environmental engineering. Over the past twenty-five years, ABC has provided engineering services to over200 municipal clients, and we are currently performing work for over 50 Authorities and Municipalities. Our varied experience includes planning, design, operations, permitting, financing and construction administration of municipal facilities. The five owners of the firm are all professional engineers and control the day-to-day operation of the company. Our staff numbers 70 individuals and includes 20 engineers, a drafting department, and two surveying crews headed by registered Professional Land Surveyors. Ten of our engineers are registered in the Commonwealth of Pennsylvania with many holding registrations in other states throughout the Mid-Atlantic region.*

114 words

Queries: Does the topic sentence express an attitude and is it paid off? Did you notice the smothered verb in line 3, the thoughts that are out of sequence in lines 4 and 5, a series that is out of parallel in line 6, and a disunity in line 7? Tip: Complete a one-word outline and see how the major topics bounce around, then chunk and rewrite.

- *OUR MISSION*

It is our mission to provide our clients and the communities they serve with the highest standard of experience, the leading edge in design innovation, and the clearest solutions in water and wastewater management ... from concept to reality.

(39 words)

Comments: The head and first sentence are redundant, and the first sentence starts with *It is ...* which should be avoided whenever possible. And *experience* cannot have a standard, can it? My rewrite:

OUR MISSION is to serve, from initial concepts through reality, our clients and their communities with innovative designs and clear solutions in water and wastewater management.

(26 words)

From another brochure

- *Our mission: To make our client's vision a reality.*

(10 words)

Comments: Wordy, lazy verb, and implies that this firm has only one client with one vision. My rewrite:

Our mission is to convert clients' visions to realities.

(Nine words, plus a sense of action for all clients.)

- *Our diverse staff includes engineers, architects, environmental scientists, surveyors, technical designers, and administrative team members.*

Comments: Parallelism be damned; you rewrite.

From a paper written by a civil engineer

- *The site is located at the top of a hill. As wind approaches a hill, it is compressed. As it reaches the top of the hill, the pressure drops and the velocity of the wind increases, making the location a more effective site for a wind turbine.*

(47 words)

Comments: Choppy, wordy, a lazy verb, more effective than what?

As it approaches the hill on which the turbine is located, the wind is compressed; then, as it approaches the hilltop, the wind's pressure increases and then decreases while its velocity increases. As a result, the site is suitable for a wind turbine.

(32 words)

- *There are two options for wind-generated energy storage: grid connected installation and stand-alone installation.*

14 words

Comments: Avoid starting a sentence with *There are* ... And why say installation twice? My rewrites: which do you like?

Wind-generated energy storage can be grid-connected or stand-alone.

(9 words (hyphenated words are counted as one.)

Wind- generated energy can be connected to the grid or not.

(11 words)

Energy generated by wind can be connected to the grid or not.

(12 words)

From another brochure

- Your vision is unique. That is why it deserves unique attention.

(11words)

Comments: Avoid starting with *that, these, those, and the same thought with which the previous sentence ended.* Instead, combine the two sentences into one:

Your unique vision deserves unique attention.

(6 words)

- *The team of professionals in our XXX section bring a broad range of experience managing projects ...*

Comment: *team* is singular in the USA; change the verb. Watch agreements when the noun and verb are separated by phrases.

- *Our reputable Water Resources Section consists of professionals specializing in water resources engineering.*

(15 words)

Comment: Redundant; what kind of specialists would you expect? Shorten to:

Our reputable Water Resources section is staffed by specialists.

(9 words)

From a cover letter to a proposal

- *XYZ is capable of providing all requested services in the RFP in-house, ultimately resulting in a cost reduction to (customer).*

(20 words)

Comments: Two smothered verbs, creating wordiness. My rewrites:

XYZ, with its in-house staff, can meet all requirements in the RFP, lowering costs.
(15 words)

XYZ's in-house staff can meet all requirements of the RFP, lowering costs.

(12 words)

XYZ can lower costs by completing all tasks with its in-house staff.

(12 words)

From a proposal

- *ABC's engineers, scientists, technical and management staff has extensive training and experience.*

(12 words)

Comment: Out of parallel, noun/verb disagreement, and two smothered verbs.

ABC's engineers, scientists, and managers are experienced and trained extensively.

(10 words)

- *Value: We strive to provide superior value in the delivery of our services.*

Comment: *We strive …* doesn't mean that we provide, and does the firm add value to its delivery or its services? You rewrite.

From a website

- *Mitigation of past, current and future risks are achieved not only by applying our knowledge of environmental policies and regulations, project experiences, and application of science and technology, but by our understanding our client's business objectives.*

(36 words)

Comment: Noun and verb do not agree. Passive and cumbersome: be active, and avoid ...*not only* ...*but by* syntax. Here's how:

ABC's engineers mitigate past, current, and future risks via knowledge of environmental policies and regulations, project experiences, application of science and technology, and understanding of clients' business objectives.

(28 words)

- *ABC has assisted clients in a variety of areas including site selection, facility decommissioning, and site redevelopment.*

(17 words)

Comment: an obvious disunity with a series that meanders from *site* to *decommissioning* and back to *site.* Easy to fix:

ABC has helped clients with site selection/redevelopment, and facility decommissioning.

(13 words)

From the Pittsburgh Post-Gazette

- *It is no secret that the Pittsburgh region is an area of slow growth. Of 15 cities measured by the Post-Gazette in its "Benchmark Series", the Pittsburgh economy is rated 12th or two rungs from the bottom.*

(37 words)

Comment: So many sins, so little time: *If it's not a secret, why tell me? It is ...region or area?...two rungs from the bottom?... a smothered verb...* so ...

The Pittsburgh region is growing slowly: our economy is ranked 12th of the 15 cities measured in our Benchmark series, a mere three rungs from the bottom.

(29 words)

The Pittsburgh region's economy is growing so slowly that it is ranked 12th of 15 cities by our Benchmark Series.

(20 words)

From a report

- *The purpose of this report is to assess technical merit and market potential in order to determine whether or not to proceed with the pursuit of greater opportunities for EPA 04-011 Method for NOX Adsorber Desulfation in a Multi-Path Exhaust System. An inventor interview was conducted on March third 2010 with a follow-up phone call on April 22, 2010. An industry expert from the Manufacturers of Emission Controls Association (MECA) was interviewed on April 25, 2010.*

(77 words)

Run-ons, and passive. My rewrite:

PURPOSE/ METHODOLOGY: Assess the technical feasibility and market potential. Then, based on our findings, elect to pursue or abandon opportunities for EPA 04-011 Method for NOX Desulfurization in a multi-path exhaust system. To help meet the purpose, we interviewed an inventor on March 3, 2010, called him for additional information on April 22, and interviewed an expert from the Manufacturers of Emission Controls Association (MECA) on April 25.

(67 words)

A few quickies

- *Lead in drinking water samples were collected by ABC at all potable water sources.*

(15 words)

Comment: Noun-verb disagreement, passive, and *drinking* and *potable* are redundant.

ABC collected samples of drinking water at all sources, then analyzed them for lead contamination.

(15 words)

- *ESCO tray scrubbers are innovative scrubbers that incorporate barriers to split the gas steam and effect the removal of soluble gases.*

(22 words)

Comment: Redundant *scrubbers ... scrubbers;* passive; disconnected thoughts; the point—barriers—is buried. Recast the sentence:

The barriers in ESCO's innovative tray scrubbers split the gas stream to enable soluble gases to be removed. (...to enable removal of soluble gases...?)

(19 words)

- *The exhaust emissions market is a wide- open field and there are many companies and institutions working on a solution to meet EPA regulations that began in 2007 with full implementation by 2010.*

(34 words)

Comment: The first four words convey the wrong meaning; run-on; a *market* or a *field*? Avoid ...*there are* ...thoughts are not connected with transitions. Resolve these sins with:

The market for minimizing exhaust emissions is vast and wide open. As a result, many companies and institutions are developing technologies that meet the EPA regulations that went into effect in 2007 and were fully implemented in 2010.

(38 words, more than the original but still more concise --remember our definition).

- *Cyclones are simple yet effective devices with no moving parts and constructed of plastic or metal materials.*

(18 words)

Comment: Disconnected thoughts; *material* is redundant, *effective* not supported. Recast the sentence:

Cyclones are simple, with no moving parts, and durable, built of plastic or metal.

(14 words)

- *As previously described, two criteria exist for the determination of lead in paint.*

(12 words)

Comment: passive; redundant.

As described previously, 'lead paint' is defined by two criteria:

(10 words)

- *Analysis of water samples collected by ABC were performed by XYZ. Samples were analyzed in accordance with EPA method XXX utilizing Atomic Adsorption Spectrometer.*

(25 words)

Comment: Passive; smothered verb. Rewrite:

Water samples collected by ABC were analyzed by XYZ according to EPA method XXX utilizing an Atomic Adsorption Spectrometer.

(19 words)

- *ABC's Technical Advisors team will be lead by VP Tom Jones , who brings 30 years experience to the project.*

(21 words)

Comment: Passive, *lead* is misspelled. Examine the two rewrites:

VP Tom Jones, with 30 years experience encompassing 22 similar projects, will lead ABC's team of technical advisors.

(17 words)

VP Tom Jones will lead ABC's team of technical advisors; he has worked on 22 similar projects during his 30-year career.

(19 words; note the specificity with the number of projects.)

- *Our Civil and Site Development Engineering Services Group provides services to real estate developers, architects, planners, industries, educational institutions, health care facilities, and contractors.*

(25 words)

Comment: Smothered verb; series out of parallel and sequence.

Our Civil and Site Development engineers serve planners, developers, architects, and contractors in industry, health care, and education.

(17 words)

PART II & WORKBOOK

ABOUT THIS SECTION

Part II of *The Power of Writing Well* extends and digs deeper into the fundamentals presented in Part I that I know are most important. The additional information will appeal to the many ambitious and motivated practitioners who are convinced that their careers and happiness will be enhanced by further sharpening their abilities to send and receive messages, and to think more analytically.

Most, if not all, architects of any structure, engineers of any discipline, constructors in any market, and businesspersons practicing in any industry were educated predominantly in the one or few skills that pertain to their profession. These same people sense and actively acknowledge that they need to develop additional skills if they and their firms are to grow and prosper. For example, they know that they face increasing demands for intelligent and relevant responses to multiple and overlapping environmental, social, and ethical issues, aka institutional reforms.

Such responses start with analytical receivings of messages from others—the function of active listening and studying-- that is followed by thoughtful sendings to target receivers—the function of clear, concise, and purposeful writing and /or speaking.

In essence, technical skills must be complemented by communications and thinking skills--and the two are inseparable.

Pete 11/21/13 3:15 PM

Comment: The Wall Street Journal reported in a special section on MBA programs that " ... employees may get by on their technical and quantitative skills for the first couple years out of school. But soon, leadership and communications skills come to the fore in distinguishing the managers whose careers really take off.'

I. CONNECTING WRITING WELL TO THE FULFILLED LIFE

Writing well is far more than putting words on a paper or screen; it is the route to enriching every human interaction, including and especially speaking, and it's a skill that everyone can learn.

But not before we reject the sad truth that in today's society we have neglected the habit of writing clear, concise, purposeful prose. Perhaps the main reason for our neglect is our addiction to so many other media that have replaced good writing, led by the ease of voice communications via telephone, Skype, and voice-recognition software that translates our ramblings to text that is laden with grammatical and syntactical meanderings that our fifth-grade teachers would never condone. Computers and smart phones have blessed shorthand, now called texting, and emails that are "good enough". All have created the feeling, the conviction, that careful crafting of words is no longer necessary.

It is, as any person can attest who has lost a contract because of a murky proposal ... lost an opportunity for employment because of an indecipherable resume ... been sued because of a muddy, ambiguous sentence in a proposal or contract ... contested an indecipherable will ... lost a friend or even a spouse because of an insensitive email ... and so on.

The negative consequences of bad writing dig their destructive tentacles into our business, financial, and personal lives. The positive consequences of good writing are equally endemic.

Yes, people are still hired because they can craft language that reflects intelligence, simply because many of us in this age of information sell nothing but our intelligence. Contracts are won because the proposal can be evaluated precisely and fairly, friendships are created because people find common ground via their language, and lawsuits are avoided when contractual obligations are stated clearly.

Pete 11/18/13 1:04 PM

Comment: A VP of a major engineering firm hires new engineers first because they can write well, second because they engineer well. "I can teach them how to be better engineers, but I don't know how to teach them to be better writers," he reasons.

The benefits of good writing continue and extend well beyond transmitting information.

Pete 7/23/13 12:19 PM

Comment: Norman Shadle reported in his book, *The Art of Communication*, that "Good communication writing pays off in both satisfaction and success. Its rewards far outweigh its achievement costs."

Good writing *creates* information, and, therefore, good writing creates intelligence and literally forces that elusive human talent that we have labeled "creativity".

The methodical, evolving process of writing actually forces new thoughts to emerge from our minds, allowing us to make sense of our surroundings, our lives, and, on a smaller scale, the document that we are composing at the moment.

Good writing gives our minds a disciplined means of expression and conjuring up the great idea that separates the ordinary from the extraordinary. It is a way to discover what we are thinking.

II. DRILLING INTO THE MAJOR TOOLS THAT DISTINGUISH GOOD SENDERS AND RECEIVERS

Every writer and teacher of writing holds dear his/her favorite magic bullets that each is convinced will create good senders and receivers.

Mine are captured in eight major tools. They reflect my many decades as a professional writer for businesses, professor of writing at two universities, and teacher/coach to top execs. I know that they are powerful, and I know that anyone can learn, apply, and benefit from them.

Pete 11/19/13 4:07 PM
Comment: I've ignored such basics as punctuation and sentence structure; astute senders already know them, and, if more is needed, they can refer to any writer's handbook or Part I of this book.

Applying one or two will yield some benefit; applying all will yield improvements that writers and their peers will notice and applaud. Guaranteed.

- **Logical reasoning:** Most business writing is structured along one of two lines of reasoning:

  22. Deductive, in which thoughts move from the more general idea or premise which is believed to be true to the specific examples, test/survey results and so on that support it, much like books, proposals, reports, and paragraphs are structured. A short example: *Carelessness killed thousands of fish and an untold number of other wildlife in the Missouri River last night. The cause has been traced to an operator at the local steel plant who failed to close a valve that would have prevented an estimated ten thousand gallons of deadly phenol from entering sewer lines that discharge into the river.* Deductive is far more prevalent than ...

  23. Inductive, in which thoughts move from specific facts or experiences to a broader, more general conclusion. Many engineers I know follow this line because, as several explained to me, they were taught it in college. Basically, they defend their premise before stating it, which is OK when the premise needs considerable defending. The same example above in inductive merely reverses the order of the sentences: *An operator at the local steel plant filed to close a valve that would have prevented an estimated ten thousand gallons of phenol to be discharged into the Missouri River. As a result, thousands of fish and an untold number of other wildlife were killed.*

- **Analytical receiving:** Analytical receivers detect far more than errors in punctuation, syntax, and such structural sins as incoherence; they also detect sendings from business that are

Pete 7/23/13 1:38 PM
Comment: University professors and other pundits tend to use *'critical thinking'* to express the same ides. I don't: *critical* implies negative or pejorative when it could be positive and constructive.

rife with sophism. Cynics say that ALL sendings from business are sophistry since all are intended to influence receivers in ways that play to the self-interest of business (e.g. higher profits), and most receivers are too unaware to discern that they are being swayed in ways that could be destructive. So, who's to blame for sophistry: manipulative, exploitive senders or accepting, unaware receivers? And what is the connection between sophism and good writing?

Pete 6/13/13 2:37 PM
Comment: False reasoning that deceives, whether intentionally (by a smart and manipulative sender) or accidentally (by a less smart or honest sender). Advertisements, brochures, and other marketing materials lead the way in sophism, but proposals, reports and so on—also marketing materials--are good examples as well.

Pete 6/13/13 2:37 PM
Comment: I think I've asked the unanswerable, a form of fallacious thinking called *evasion* or *hedging*.

Just this: To be most influential, good writing must avoid the fallacies of thinking that lead to sophism, and, in turn, disbelief in your message and you as a writer. Fallacies are detected by astute people either consciously or intuitively; they are part of most groups of receivers and must be considered when developing the message. (The rules change if the sending is intended solely for the obtuse, which in business is rare outside of advertising.)

Pete 6/13/13 2:37 PM
Comment: Note the two-word sentence that isn't a sentence but is a workable transition to the next paragraph.

Scholars have identified and named at least 44 fallacies of thinking that permeate our communications. I've selected five that I feel are most prevalent in business:

51 Overblown generalizations that are unsupported and, therefore, faulty and unbelievable.

Pete 6/13/13 2:37 PM
Comment: Aka *sweeping generalizations* or *absolute statements* that are thought by some pundits as impossible to support.

Example: *ABC Engineers was built on the philosophy of getting the job done right the first time,* is an inadequate topic sentence that is followed by another overblown generalization: *We've developed a reputation for the completeness of our permitting and design documents, which expedite implementation of complex projects.*

Pete 6/13/13 2:37 PM
Comment: On top of the fallacious thinking, the noun and verb do not agree. Ugh. And the meaning is wrong: Can a reputation expedite?

Good writers support overblown generalizations by asking and answering *why do I know that to be true? How can I prove the point?* Asking and answering why and how several times---some say at least seven—guarantees complete and believable support.

52 *Non sequitur,* aka it *doesn't follow.* I discussed this as part of cohesion, and it's worth revisiting since it's so common. See the example in the bullet above; it's perfect since the first sentence is totally unrelated to the second.

53 Appeal to experience. Skilled spinners often imply that they are more competent simply because they are more experienced—a big jump in logic that can apply to age as well. Analytical receivers ask: what were the results of that experience? How do those results apply to me?

Example: For 25 years, ABC has provided innovative services to our clients, many of whom have been with us since our inception.

> Pete 11/18/13 1:05 PM
> **Comment:** Is "our" needed? Would ABC serve other clients that weren't theirs?

> Pete 6/13/13 2:37 PM
> **Comment:** *Many of* can be one or 99 percent, and the receiver must guess which. Be specific. Repeat business can be a powerful point for sales and marketing, but not when it is vague.

26 Confusing cause/effect, often related to *non-sequitur*. See the example in the first bullet. Completeness of documents does not necessarily expedite projects; many steps are missing that might or might not connect the two thoughts.

27 Ignoring your and other persons' biases. Avoid bias in your sendings and detect them in receivings by keeping an open mind, delaying judgments, and, above all, considering the source. Good examples of bias are, again, ads, followed closely by public relations releases; although they are published in newspapers and trade magazines as news, many are in fact the most blatant sophisms.

> Pete 6/13/13 2:37 PM
> **Comment:** Read the letters to editors in any newspaper and you will read perhaps the most blatant biases. Consider the source.

- **Empathy with receivers:** *Empathy* is that power to understand another person's needs, wants, opinions, and points of view. Senders can develop empathy with receivers by fantasizing how receivers will react to every sending's content (what is said), structure (the order in which the content is presented), and tone (the choice of words).

> Pete 6/13/13 2:37 PM
> **Comment:** Tone conveys attitude or mood, which can be, among other emotions, serious or flippant, respectful or condescending, friendly or hostile, intimate or detached, and so on. Tone tells receivers how senders feel about the subject and must be compatible with purpose.

The objective is to send the correct number of suitable words/thoughts (concision) that are understood (clarity) and meet, as precisely as possible, the purposes of both senders and receivers (purposeful, or on-point).

> Pete 7/23/13 4:00 PM
> **Comment:** Sendings that do not play to the needs and wants of receivers and instead play to the needs and wants of senders are called solipsistic or narcissistic. Regardless of its label, such sendings fail to meet anyone's purpose (s). Empathy with receivers helps senders to avoid that trap.

Receivers react to sendings that do not meet these criteria with confusion, frustration, and downright hostility that transcends annoyance. They then avoid careful receiving by skip-reading or day-dreaming, and senders never get their messages across. The sad result is lower productivity (aka wasted time), lost contracts (aka wasted money and fewer employees), damaged relationships (aka unhappiness and lost customers), and more.

Careless senders suffer needlessly as well; they are labeled "not too bright", just what they don't want to be labeled, and their careers are stalled, perhaps irreparably.

- **Cohesion/unity/sequence:** Sentences, paragraphs, and entire documents are cohesive when they flow smoothly and clearly toward a defined conclusion or direction that is set by senders' and receivers' **purpose(s).** All the sentences are connected logically, and every word, phrase, and sentence explains, demonstrates, illustrates, and clarifies the purpose (s) of the piece to create a unified whole. To examine the other side of the same coin, the sender never strays from the purpose(s), never takes an extraneous excursion, and all words and thoughts that do not add to the story line—yes, business sendings have a story line—are deleted by the careful sender. Many such excursions are **disunities**— breaks in the logical flow of ideas—and they destroy both clarity and concision, annoying receivers with irrelevant blather as they do so. See example 6 for the perfect disunity.

Other excursions are ***non sequiturs***—basically a sentence or longer piece in which the opening assertion or premise is followed by an unrelated thought, breaking the logical sequence. Careful receivers often say "it doesn't follow" instead of "it's a *non sequitur*".

A perfect and frequent example in business: *ABC is an industry leader in the development and implementation of solutions. Our client driven methods provide time and cost savings through the streamlining of key business processes.* The first sentence—our leading position-- is totally unrelated to the second—we save time and money.

Pete 11/23/13 12:02 PM
Comment: This sentence is loaded with other sins as well, starting with passive voice, smothered verbs, and a missing hyphen. To revise: ABC leads the industry by developing and implementing solutions. Client-driven methods and efficient business processes compress schedules and reduce costs.

- **Active and passive voices:** We tend to use active voice in every day speech and writing: "Mark designed a beautiful bridge." (5 words) not "A beautiful bridge was designed by Mark."(7 words); "Josh negotiated a lucrative contract", not "A lucrative contract was negotiated by Josh." The actor in each active sentence is the subject of the sentence and is obvious; the actor in each passive sentence is less obvious and is not the subject of the sentence, which is acted upon.

Active voice typically needs 20-30 percent fewer words to express the same thought while adding clarity, concision, and a sense of action.

Nevertheless, and despite these obvious advantages of active voice, passive has its legitimate uses: 1. when the actor is unknown or irrelevant to the message (note that I used passive to describe passive. I could have said: When the sender doesn't know the actor or considers him to be unimportant.); 2. when the object being acted upon is more important than the actor: *Our CEO was fired by the board for poor planning* presumes that the CEO is more important than the board. *The board fired the CEO for poor planning* assumes the opposite.

Pete 6/13/13 2:37 PM
Comment: I demonstrated a numbered list in a horizontal format; I could have demonstrated a bulleted or numbered list in a vertical format just as easily.

Scientists and engineers tend to overuse passive, in part I've been told because lawyers think that passive deflects responsibility. Not true; passive makes it more difficult to assign responsibility but it surely does not deflect it. Another reason is that many scientific and engineering sendings describe a process and the actor is either unknown or unimportant. Consider this sentence written by a chemical engineer:" It

Pete 6/13/13 2:37 PM
Comment: Starting a sentence with 'It is' automatically puts you in passive voice.

is clear to me that we are polluting the air with our current practices.'' (16 words) She could have written: "I am convinced that we are polluting the air with our current process". (13 words)

Some pundits say that the average business document should be 65% active, 35 % passive. You can check your documents using the software on most computers. The same software will highlight passives so that you can change to active as you see fit.

- **Lists:** A list can separate and highlight complex prose by moving it out of the main text and into a series of key ideas that can be read more easily. Use a numbered or alphabetized list to show a hierarchy of importance or sequence of events, such as the steps in a process; use a bulleted list otherwise. Numbered or alpha lists can be arranged horizontally (less common) or vertically (more common); bullets are always arranged vertically.

Lists, like the topic sentences of paragraphs, are used effectively to highlight important points and to support a controlling idea (generally a broad or abstract idea that demands further explanation before it becomes an overblown generalization.

Caveat: Limit a list to three or four points to be more understandable and memorable. Telephone and social security numbers are arranged in threes and fours for that reason, as are auto license plates, when the first three numbers or letters are separated by a symbol and followed by four more numbers and/or letters.

More caveats: a) Be parallel by always starting a list with the same word form; b) Be stingy by using lists only when needed since too many lists create a choppy document that is difficult to read and appears to be more of an outline than prose; and c) be aware that you can always reverse a list to prose and still benefit from the exercise.

Pete 6/13/13 2:37 PM
Comment: I demonstrated a horizontal list where I could just as easily demonstrated a vertical.

- **Parallelism:** Using similar grammatical constructions to reinforce similar ideas adds to coherence of sentences, paragraphs, lists, outlines such as tables of contents, and entire documents. Be parallel when writing a) items in a series: Eat, drink, and be merry; b) paired items: Your memo was short and sweet, and c) in lists: We shopped for bread, butter, and cream. Faulty parallelism abounds in the examples that follow, and most, it seems, are in lists that start with different forms of words, so: If you start an item in a list with a verb or other form, start all items with the same form. Refer to example 9.

- **Unified paragraphs:** A paragraph is a group of related sentences, and it can stand alone as a complete piece or be part of a longer document. Begin a new paragraph to move from one topic/location/time period to another, and introduce a new step in a process. Regardless, adhere to the guidelines for cohesion, parallelism, and punctuation.

The typical paragraph is crafted in three parts:

a) The topic sentence states the main idea to help receivers follow the discussion; it is almost always the first sentence in the paragraph to help receivers understand the main idea immediately and allow it to be developed using deductive reasoning.

Pete 6/13/13 2:37 PM
Comment: Topic sentences are rarely placed in the middle or end of a paragraph.

b) Support of the topic sentence in the form of examples, statistics, and expert testimony---all of which are acceptable to receivers only if that are accurate, unbiased, sufficient, and relevant. The topic sentence is an *overblown generalization* if its support fails to meet one or more of these criteria.

Pete 6/13/13 2:37 PM
Comment: See the discussion on analytical receiving below.

c) A conclusion and/or transition to the next paragraph.

The sections of a unified paragraph are connected smoothly by transition words and phrases that signal receivers that a new thought is on its way. Beginning senders tend to ignore transitions, feeling, erroneously, that they add unnecessary and flowery words. Not so. A few suggestions:

To add a thought, use *besides, what's more, furthermore, in addition, for example, for instance.*

To add emphasis, use *in fact, indeed, therefore, consequently.*

To grant an exception, use *to be sure, of course, though, still, however, on the other hand, nevertheless.*

To arrange ideas in order, say *first, second, next, then, finally.*

To sum up several thoughts, say *in short, in brief, in summary, in conclusion.*

Analytical receivers detect good sendings when concepts are supported by facts and relevant, solid evidence instead of opinions or bias. They also detect fallacious reasoning and other attempts to spin the message.

Analytical receivers detect these and other flaws not only in the sendings of others, but also in their own. For example, I know that this book is biased in favor of the tools that I know are valuable for my target receivers, an opinion that I'm certain is not shared by all writers and teachers, as I point out in the introduction to this chapter.

III. APPLYING THE MAJOR TOOLS TO REAL-WORLD EXAMPLES

I am besieged with examples of prose that violate the most basic guidelines for logic, clarity, concision, and purposefulness. I've selected eleven; *the originals are italicized* and include comments, the corrected in regular type.

I've also added a short receiver profile and purpose for each to help analyze the appropriateness of the piece.

1. A national AEC firm describes its business

Receivers: clients and prospects, all knowledgeable of the industry and its jargon.
Purpose: marketing/sales: influence receivers that ABC's local presence and national network positions ABC as the best choice for a specific project.

LOCAL PRESENCE, NATIONAL STRENGTH

ABC's local office is on Easy Street in Potown PA. Our new facility has an accredited full-service laboratory and is centrally located to provide great customer service to all our clients. Our local staff of 75, provides a depth of knowledge and resources to be responsive to all of our clients needs.

ABC has a network of approximately 125 offices located across North America. When our local office works on a project, you are assured of getting the resources of a national company. It is the perfect combination of local expertise on technical matters, a working familiarity with members of the regional business and regulatory communities and quick, accurate, economical response that only ABC can give you.

SINGLE SOURCE FOR ALL YOUR NEEDS

- A. *Certified Professional Staff*
- B. *Responsiveness*
- C. *Values*

PROVEN EXPERIENCE

- ➢ *Stadiums and Arenas*
- ➢ *Federal, State, Municipal Facilities*
- ➢ *Educational Facilities*
- ➢ *Automotive Industry*
- ➢ *Financial Institutions*
- ➢ *Airports and other major Transportation Facilities*

<u>We have Large-Scale National Resources to Support our Local Offices</u>

Pete 6/13/13 2:37 PM
Comment: A throwaway; combine the thought with others. Where's the attitude?

Pete 6/13/13 2:37 PM
Comment: No comma

Pete 6/13/13 2:37 PM
Comment: Pity the poor apostrophe.

Pete 6/13/13 2:37 PM
Comment: Another throwaway.

Pete 6/13/13 2:37 PM
Comment: Doesn't this contradict the last sentence in the first paragraph?

Pete 7/23/13 4:13 PM
Comment: The bullets do not support the head, a perfect disunity.

Pete 6/13/13 2:37 PM
Comment: The bullets are not parallel and they are overblown generalizations.

Pete 7/23/13 4:14 PM
Comment: The bullets do not support the head.

Pete 6/13/13 2:37 PM
Comment: The bullets are not parallel.

Pete 6/13/13 2:37 PM
Comment: Redundant and contradictory to the second sentence in the second paragraph.

PROJECT TEAM

ABC commits its significant technical resources to the XXX Venue Group, LLC and their design consortium, which includes YYY.

ABC's Technical Advisors team will be lead by Vice President Mr. Sam Spade, P.E., who brings 30 years experience and Chief Engineer Jim Jones .P.E., who brings 20 years of experience to the project. Representative ABC related project lead by Mr. Spade include: BBB, CCC and EEE. Representative ABC projects lead by Mr. Jones include: FFF, GGG and HHH. Providing technical support on this project will be Mr. Ken Kenny, Ph.D., P.E., Mr. Lou Lenny, P.E., and Mr. Gary Golly, P.E.

Pete 6/13/13 2:37 PM
Comment: Unrelated to the head.

Pete 6/13/13 2:37 PM
Comment: Sic.

Pete 6/13/13 2:37 PM
Comment: Sic again.

Pete 6/13/13 2:37 PM
Comment: Ho hum: How many times can you say P.E. and Mr? Move P.E. to the top of a list

Two hundred twenty words of incoherent, unparallel, redundant, and ungrammatical irrelevance without a single benefit for the target receivers.

ABC's new, local office on Easy Street in Potown PA is centrally located to help deliver superior, rapid, and responsive service to clients throughout Western PA. The building houses a full-service laboratory and 75 engineers and scientists who have proven their competence on hundreds of projects. They are complemented by X thousand peers headquartered at 125 similar facilities throughout North America. Clients benefit by working with a single source for all services, helping to reduce paper work and administrative costs.

Pete 6/13/13 2:37 PM
Comment: A map?

ABC's team of technical advisors, all P.E.s, will be led by:
Messrs Sam Spade, with 30 years experience on similar projects, including BBB, CCC, and EEE; and Jim Jones, with 20 years experience including FFF, GGG, and , and HHH.

The team will be supported by:
Messrs Ken Kenny, Ph.D., Lou Lenny, and Gary Golly.

137 Words

2. An author/professor describes her book

Receivers: Other professors, college administrators
Purpose: Help with planning courses and encourage receivers to buy her book

My recent book, VVV, provides a full description of an integrated approach to designing college courses. This paper outlines the key ideas and components of this model.

Pete 6/13/13 2:37 PM
Comment: Smothered; watch for *provide*—it is usually followed by a noun, in this case *description*, that can be a verb, *describes*.

Pete 6/13/13 2:37 PM
Comment: Try to not start a sentence with this, that, those, these. Instead, combine it with the preceding sentence.

27 words

My recent book, VVV, describes an integrated approach to designing college courses; key concepts are described herein.

17 words

OR: I describe herein the key concepts for an integrated approach to designing college courses. For a full description, please see my recent book, VVV.

25 words

3. A consultant confirms the scope of work for an asbestos survey

Receivers: Other environmental scientists
Purpose: Assure that the scope is clear to all parties

> *Mr. HJK of XYZ Authority retained ABC to conduct an asbestos survey to estimate the type, location, and quantity of suspect Asbestos Containing Material (ACM) on the lower Building one and Building Two roofs at the Painful Shopping Center. As directed, ABC performed the work under an existing Agreement for environmental services between XYZ and PDQ.*

56 words

Mr. HJK of XYZ Authority retained ABC to survey the roofs of Buildings One and Two at the Painful Shopping Center. Its purpose: to estimate the type, location, and quantity of suspect Asbestos-Containing Material. As directed, the work was completed under an existing agreement for environmental services between XYZ and PDQ.

51 words

4. A consultant explains the procedure for an asbestos survey

Receivers: other consultants.
Purpose: To detail how the work was completed

Prior to sampling, a bulk material sampling plan was established. Representative asbestos bulk samples of suspect asbestos materials were collected, containerized, labeled and submitted to the laboratory for PLM analysis with further instructions to include Point Count analysis if the PLM result was found to be between 1 and 10 percent asbestos.

Appendix A of this report contains a table, which summarizes the materials sampled, and quantifies those that tested positive for ACM.

Appendix B of this report contains copies of the original laboratory asbestos analytical test reports for the materials sampled and tested.

Appendix C of this report contains the Asbestos Bulk Sampling Forms, which document the conditions and amounts of the sampled materials.

115 words

Before sampling, ABC planned the work as follows:

- Collect, containerize, and label representative samples of suspect materials;
- Submit the samples to the laboratory for Point Count analyses and PLM analyses of samples with between 1 and ten percent asbestos.

Please refer to *Appendix A* for a summary of the materials tested, including those with ACM; *B* for copies of the original laboratory reports; and *C* for sampling forms and the conditions and amounts of sampled materials.

78 words

5. An engineering firm describes its scope of work

Receivers: other engineers.
Purpose: to clearly define the tasks for an underground survey.

Our scope of work will be:

- *Field and Laboratory Testing*
- *Geotechnical Design Criteria for Foundations*
- *Geotechnical Design for Slabs*
- *Geotechnical Design for Pavement systems*
- *Groundwater conditions as Pertain to the Proposed Construction*
- *Geotechnical Factors that may Impact Construction Procedures*
- *Foundation Settlement Based on Conditions Encountered*
- *Geotechnical Recommendations for Retention Systems*
- *Geotechnical Recommendation Pertaining to Underpinning*
- *Seismic Site Classification*
- *Site Specific Seismic Ground Motion Analysis*

Pete 6/13/13 2:37 PM
Comment: Attitude?

Pete 6/13/13 2:37 PM
Comment: Eleven incomprehensible bullets. Notice how the topics bounce around from Testing, Geotechnical, geotechnical, Geotechnical, Groundwater, Geotechnical, Foundation, Geotechnical, Geotechnical, Seismic, and Seismic---the perfect disunity.

75 words

Our scope of work will be comprehensive and thorough, and will encompass:

7. Geotechnical engineering:

a. define design criteria for foundations, slabs, and pavements;
b. ---recommend design criteria (?) for retention systems and underpinning; and
c. ---identify factors that may impact construction such as groundwater and other underground conditions.

8. Seismic studies: classify the site and analyze ground motion.

9. Testing in the field and laboratory.

1 words

6. A consultant describes its business

Receivers: Other consultants and buyers of consulting services, all knowledgeable.
Purpose: Marketing

INTRODUCTION

CDE Environmental and Remediation, Inc. (CDE E&R)is a full service environmental consulting company with offices in both Eastern (Potown) and Western (Wotown) Pennsylvania. We offer responsive, reliable expertise to industry and government to provide cost-effective solutions to environmental problems. Our experience and dedication ensures that our clients' expectations are met or exceeded.

CDE's executive management has fifty years of combined experience in environmental consulting.

The Company is a subsidiary of XYZ, a privately owned innovator in the manufacturing of pigment dispersions and color concentrates. XYZ supplies colorants to major producers of plastics, printing ink, cosmetics and coatings.

91 words

I honestly can't rewrite this useless blather without assuming what this firm does and why it's competent. But I'll try:

CDE is the choice for responsive, reliable expertise needed to resolve a wide range of environmental problems faced by industry and government. The firm's expertise encompasses:

26 words

Oddly enough, the carnage of the language continues, and is repeated, on the same page as the INTRODUCTION.

Pete 11/21/13 3:18 PM
Comment: Did I change the meaning?

Pete 6/13/13 2:37 PM
Comment: The firm's name is out of parallel, which hints of the blips on the horizon.

Pete 6/13/13 2:37 PM
Comment: Needs a hyphen.

Pete 6/13/13 2:37 PM
Comment: Ho hum: no attitude.

Pete 6/13/13 2:37 PM
Comment: Aha! One person with 50 years experience? Two with 25 years each? Fifty with one year each? Are you impressed?

Pete 6/13/13 2:37 PM
Comment: Capital?

Pete 6/13/13 2:37 PM
Comment: This entire paragraph is the perfect disunity; receivers do not care. If it must be said—and I see no reason other than to satisfy some corporate ego-- make it a footnote.

Pete 6/13/13 2:37 PM
Comment: See 7 below.

WHY CDE E&R?

Pete 6/13/13 2:37 PM
Comment: Never, ever ask a question in a headline.

CDE E&R's mission is to provide our clients with reliable, cost-effective solutions to their environmental problems.

Pete 6/13/13 2:37 PM
Comment: Said in INTRO; delete here.

Our capable, seasoned professionals apply strong management skills and technical expertise to ensure your projects are well planned, fully executed and thoroughly documented on time and within budget.

Pete 6/13/13 2:37 PM
Comment: Said in INTRO; delete here.

Our professional staff combines expert judgment, state-of-the-art resources and a company-wide dedication to understanding the needs of our clients to provide superior service.

Pete 11/23/13 3:02 PM
Comment: Outdated cliché. Delete.

Pete 6/13/13 2:37 PM
Comment: Combine the second and third paragraphs.

Seventy empty, hollow words that repeat much of what was said in the INTRO. Again, I'll try to rewrite if you'll remember that I have very little substance to work with.

COUNT ON CDR TO CONTROL COSTS AND ACCELERATE YOUR PROJECTS

Our seasoned professionals apply, to every project:

- Strong management skills and technical expertise supported by the latest resources;
- Expert judgment based on years of experience; and
- Deep and thorough understanding of clients' needs.

The results: Projects that met or exceed clients' performance and cost expectations.

Pete 7/23/13 4:24 PM
Comment: I have no idea what that means: Computers? Copiers? Can't be people; they aren't "latest".

Fifty-eight words that at least have a bit of substance and sell.

I've saved the worst for last in this marketing piece: The description of what this firm says it does for its existence. Read it and weep:

SERVICE AREAS

Pete 6/13/13 2:37 PM
Comment: A head with no substance; avoid.

The environmental and engineering consulting and remediation services offered by CDE are listed below:

Pete 6/13/13 2:37 PM
Comment: Why passive voice? Why repeat the head?

- *Phase II and III Environmental Site assessments*
- *Brownfields/PA Act 2 Property Evaluations*
- *Contaminated site Investigations*
- *RAD Investigation and remediation Services*
- *Storage Tank Management*
- *Hydro geologic Investigations & Water Supply Studies*
- *Permitting*
- *Air emissions Inventories, Monitoring & Control*
- *Engineering and Consulting Services for the Mining Industry*
- *Regulatory Liaison*
- *Environmental Health and Safety Audits*

- *OSHA/Environmental Compliance Monitoring*
- *Water and Waste Water System Design and Installation*
- *Civil/Environmental Engineering*
- *Expert witness and Litigation support*
- *Landfill Cell Construction and Capping*
- *In-situ and Ex-situ Soil remediation, Waste Stream Identification, Consolidation, and Disposal*
- *Facility Decontamination, Decommissioning & Demolition*
- *Site Restoration*
- *Wetland Restoration and Streambank Stabilization*
- *NJ ISRA Services*
- *Soil and Ground Water remediation System design, Installation & Management*
- *Municipal Engineering*

Twenty-three out-of- parallel bullets and 152 jumbled, disorganized words. Ugh.

The fix? Let's start by organizing the bullets in the two services in the firm's name: Environmental and remediation, and see if all the bullets can fit: (Hint: They can't.)

- ENVIRONMENTAL
  --Phase II and II site assessments; contaminated site investigations
  --Hydrogeologic investigations and water supply studies
  --Permitting
  --Air emissions inventories, monitoring & control
  --Regulatory liaison
  --Environmental health and safety audits
  --OSHA /environmental compliance monitoring
  --Land fill construction and capping

- REMEDIATION
  --Brownfields/PA Act 2 property evaluations
  --In-situ and ex-situ soil remediation, waste steam identification, consolidation and disposal
  --Facility decontamination, decommissioning & demolition
  --Site restoration
  --Wetland restoration and streambank stabilization
  --Soil and groundwater remediation system design, installation & management

- RELATED
  --RAD investigations and remediations
  --Storage tank management

--Engineering and consulting for the mining industry
--Water and waste water system design and installation
--Civil/environmental engineering
--Expert witness and litigation support
--NJ ISRA services
--Municipal engineering

The good news is that I've pared 23 bullets to three and have chunked the bullets by service category; the bad news is that I still have 23 sub-bullets so they need to be chunked.

CHANGE THE HEAD: CDE OFFERS A FULL RANGE OF SERVICES FOR GOVERNMENT AND INDUSTRY ...OR

CDE: THE CHOICE FOR A FULL RANGE OF SERVICES

- ENVIRONMENTAL
  --Assessments/investigations: Phase I and II, contaminated sites, hydrogeologic
  --Monitoring: Air emissions, health and safety, OSHA compliance
  --Permitting
  --Construction: landfills
- REMEDIATION
  --Evaluations: Brownfields/Pa Act 2 properties, RAD
  --Restorations: sites, wetlands, stream banks
  --Design, installation, management: Soil and groundwater systems
  --Decontamination, decommissioning, demolition
- RELATED
  --Investigations: RAD
  --Management: storage tanks
  --Design and installation: water and wastewater systems
  --Engineering: civil/environmental, mining, municipal
  --Other: NJ SRA, expert witness and litigation support

I. words and three bullets and 13 sub-bullets. Is it as concise and unified as possible?

7. A consultant describes a process

Receivers: Other consultants and purchasers of consulting services
Purpose: Explain the process and demonstrate competence

The process often has many parts including:

Pete 6/13/13 2:37 PM
Comment: Where's the comma?

- *project scoping meetings with the project sponsor and relevant regulatory agencies;*
- *development of project alternatives;*
- *impact studies in a number of specialized areas leading to the preparation of an environmental document; and*
- *public information meetings and hearings.*

Pete 6/13/13 2:37 PM
Comment: Bullets two and three form a disunity and bullets one and four can be chunked.

Pete 6/13/13 2:37 PM
Comment: Bullets 1, 2, and 4 start with compound adjectives; bullet 3 with a verb.

48 words

The process often includes:

- **meeting** with the project sponsor, relevant regulatory agencies, and the public;
- **studying** specialized areas leading to an environmental document; and
- **developing** alternatives.

Pete 6/13/13 2:37 PM
Comment: All the bullets start with the same verb form.

29 words

8. A Civil Engineer describes a site

Receivers: Other Civil Engineers
Purpose: Explain site conditions and demonstrate competence

The project site and vicinity overlook the Mon River Valley in an area of significant relief with topography that varies from gently rolling hills to steep coal mining strip benches and ravines. The site includes grass-covered fairways and greens, deciduous forest-covered slopes, and residential development. The site relief varies from an elevation of 721 feet at Turtle Creek to approximately 1220 feet near the clubhouse near the top of the hill.

Pete 6/13/13 2:37 PM
Comment: Out of parallel.

Pete 6/13/13 2:37 PM
Comment: The second sentence is a disunity, and the three sentences can be combined into one.

71 words

The project site and its vicinity overlook the Mon Valley: its topography varies from gently rolling hills to steep coal -mining strip benches and ravines, its elevation varies from 721 feet at Turtle creek to 1220 feet near the clubhouse near the top of the hill, and its surface conditions vary from grass-covered fairways and greens to deciduous forests on slopes and residences.

63 words

9. A consultant describes its services

Receivers: Other civil engineers
Purpose: Display the firm's broad range of services

Pete 6/13/13 2:37 PM
Comment: Because it demonstrates the points I want to make so perfectly, I've repeated and expanded the example I used when I discussed overblown generalizations.

ABC Engineers was built on the philosophy of getting the job done right the first time. We've developed a reputation for completeness of our permitting and design documents, which expedite implementation of complex projects. Direct management by one of five ABC partners on every project maximizes quality, efficiency and accuracy.

OUR PARTNERS WILL BECOME YOUR PARTNER ...

Services:

- *Field Surveying*
- *Grading, Access and Site Design*
- *PA DEP ESCGP-1 Permits*
- *Susquehanna River Basin Permitting and Clearances*
- *Impoundment Permitting and Design*
- *PENNDOT Highway Occupancy Permits and Local Municipal Roadway Permitting*
- *Local Land Development Ordinance Submission Preparation*
- *Land acquisition/Right-of-Way Support*
- *Construction Observation*
- *Endangered/Threatened Species)PNDI) Searches*
- *Wetland Delineation, Permitting and Mitigation*
- *Stream Assessments, Relocation and Restoration*

Pete 6/13/13 2:37 PM
Comment: Overblown and without the serial comma.

Pete 6/13/13 2:37 PM
Comment: Another overblown generalization.

Pete 6/13/13 2:37 PM
Comment: A blatant violation of the 3-4 and parallelism rules described under *lists*.

One hundred twenty seven words that violate so many principles of good writing that the piece cannot be fixed—except the bullets, which, if the piece is to be cohesive, must be grouped by PERMITTING and DESIGN:

PERMITTING

- Susquehanna River Basin
- Impoundments
- PENN DOT Highway Occupancy and Local Roadways
- Local Land Development
- Wetland Delineation/Mitigation

DESIGN

- Field Surveys
- Grading, Access, and Site
- Impoundments

ASSESSMENTS AND SUPPORT

- Land acquisition / Right-of Way
- Construction Observation
- Stream Relocation and Restoration

Pete 6/13/13 2:37 PM
Comment: I could have started the sentence with *127 words* ...but that would violate the rule that numbers are spelled out when they start a sentence.

Pete 6/13/13 2:37 PM
Comment: Se the second line in the first paragraph, in which *permitting* and *design* are introduced in that order.

Pete 6/13/13 2:37 PM
Comment: Oh well, ABC didn't mean it when it talked about its permitting and design services. The firm offers much more, but receivers are forced hunt for what that may be.

10. A Letter to Request an RFP

Receivers: Other geotechnical engineers

Purpose: Explain qualifications for a contract

Dear Mr. X;

Thank you for considering ABC for the geotechnical engineering services for the new (facility).

ABC has a long, successful history of collaboration with (other firms) on major projects across the United States and offers the strength, depth and commitment of one of the largest consulting engineering firms in the United States. In addition, our YYY Division, which will lead this project, has offered geotechnical engineering in the area for more than 100 years. In fact, our YYY Division has performed a limited subsurface investigation at the proposed site, which has allowed us to tailor our proposal based on direct knowledge of existing site geology and subsurface conditions.

ABC is committed to including meaningful participation by local firms in the area. To this end, our proposal includes X% of local MBE, and Z% of local WBE Participation.

ABC's unique strengths and local knowledge of the site will result in a cost effective geotechnical solutions for the new facility. We look forward to meeting with you to further discuss our qualifications and approach for this significant project.

178 words

Dear Mr. X:

Thanks for considering ABC for the geotechnical engineering at the new (facility).

When evaluating proposals, please note that ABC offers a combination of benefits that will expedite and control the costs of the project, including:

- A #-year history of working closely and successfully with (other firms) on # similar projects;
- A knowledge of geologic and other conditions at the site gained from a previous contract for limited subsurface drilling·, enabling us to tailor this proposal to your precise needs and to accelerate the work once the contract is awarded;

· The work was completed by YYY, a division of ABC (one of the largest consulting firms in the United States) and the lead contractor for the geotechnical work for this project.

- Broad capabilities derived from more than # scientists and engineers at our (local) office; they are complemented by # others based throughout the United States;
- A commitment to hiring local MBE (% participation) and WBE (% participation) firms.

We look forward to further discussions of our qualifications, and to working with you on this and future projects.

219 words.

INDEX

ABOUT THE AUTHOR

Pete Geissler has written more than three million words as a professional writer. He has taught writing at Duquesne University, Carnegie Mellon University, the Engineers' Society of Western Pennsylvania, the Pittsburgh Technology Council, and various consulting and manufacturing companies. *The Power of Writing Well* is the second in a series which includes *The Power of Being Articulate* and *The Power of Dignity*. *The Power of Ethics* will be published in 2014. *Big Shots' BullShit* and *Divorce Can be Such Sweet Sorrow* are also going to press. Pete can be reached at www.petegeissler.com which will take you to Pete's new publishing venture, The Expressive Press.

CPSIA information can be obtained at www.ICGtesting.com
Printed in the USA
BVOW05s2005180114

342264BV00003B/10/P